Handbook of Food Engineering

Handbook of Food Engineering

Travis Summers

MURPHY & MOORE
www.murphy-moorepublishing.com

Murphy & Moore Publishing,
1 Rockefeller Plaza,
New York City, NY 10020, USA

Visit us on the World Wide Web at:
www.murphy-moorepublishing.com

ISBN: 978-1-63987-281-7 (Hardback)

Cataloging-in-Publication Data

Handbook of food engineering / Travis Summers.
p. cm.
Includes bibliographical references and index.
ISBN 978-1-63987-281-7
1. Food industry and trade. 2. Food industry and trade--Appropriate technology. 3. Food engineers.
4. Agricultural engineering. I. Summers, Travis.
TP370 .H36 2022
664--dc23

Table of Contents

Preface

Food engineering is a branch of science which applies the principles of engineering and mathematics to food manufacturing. It also includes the analysis of various food related processes such as food conservation, food packaging, and handling and distribution of food products. It is a multi-disciplinary field which employs the principles of mechanical, electrical, civil and agricultural engineering. Concepts of food microbiology, biochemistry and rheology also play an important role in the applications of food engineering. Some of the important industrial processes studied under it for producing safe and sustainable food are refrigeration, evaporation, packaging, heat transfer and food safety management systems. It also makes use of advanced monitoring and control systems to facilitate automation and flexible manufacturing of food. The topics included in this book on food engineering are of utmost significance and bound to provide incredible insights to readers. It studies, analyses and upholds the pillars of food engineering and its utmost significance in modern times. This book is a complete source of knowledge on the present status of this important field.

Given below is the chapter wise description of the book:

Chapter 1- The professional and scientific field which is involved in the application and interpretation of principles of science, mathematics and engineering to the manufacturing and other operations related to food is termed as food engineering. All the diverse aspects of food engineering such as food preservation, food processing, food fortification and food technology have been briefly introduced in this chapter.

Chapter 2- There are various principles which play a pivotal role in the preservation and processing of food. A few of them are ultrasonic processing, retort processing, heat preservation and processing, low temperature preservation and processing and Ohmic heating and inductive heating. This chapter discusses in detail these principles of food processing and preservation.

Chapter 3- The technology involved in the packaging of food for the purpose of protecting it from tampering as well as fulfilling its physical, chemical or biological requirements falls under the umbrella of food packaging technology. The different aspects of food packaging technology such as aseptic packaging, active and intelligent packaging, and packaging of dairy products have been thoroughly discussed in this chapter.

Chapter 4- Food engineering entails the study and applications of various concepts and devices such as biosensors, food adulteration and non-destructive food analysis. Biosensors find a variety of uses in food engineering, especially in detecting pathogens and food toxins. The topics elaborated in this chapter will help in gaining a better perspective about these aspects of food engineering.

Chapter 5- The foods which are sourced from organisms whose DNA has been changed through genetic engineering are known as genetically modified foods. Genetic engineering techniques enable the production of crops which are resistant to pathogens and herbicides as well as possess better nutrition profiles. This chapter closely examines the key concepts related to the advantages and disadvantages of genetically modified crops to provide an extensive understanding of the subject.

At the end, I would like to thank all those who dedicated their time and efforts for the successful completion of this book. I also wish to convey my gratitude towards my friends and family who supported me at every step.

Travis Summers

Chapter 1

Food Engineering: An Introduction

The professional and scientific field which is involved in the application and interpretation of principles of science, mathematics and engineering to the manufacturing and other operations related to food is termed as food engineering. All the diverse aspects of food engineering such as food preservation, food processing, food fortification and food technology have been briefly introduced in this chapter.

Food engineering is a multi-disciplinary field that is responsible for the production and preservation of food. The focus of a food engineer is to constantly improve the food production process, whether that is introducing energy efficient machinery, or developing a new type of packaging to increase the shelf-life of a product. Working closely with food technologists and food scientists, their combined goal is to provide a healthy, tasty, long-lasting product while also ensuring global food supplies remain plentiful and safe. This diverse role is at the forefront of innovation within the food industry.

Foods represent a complex mixture of hundreds of components, which contribute to their characteristic flavor. Nowadays, assuring the food's flavor stability represents an issue of food industry involved actors, becoming one of the main technological challenges. An educated consumer has increasing demands related to food quality perception over the last years. As a consequence, the food flavor instability is a critical quality issue of food industry. Achieving food's quality in respect to its flavor and stability is a complex process, especially by considering the large variety of flavor compounds and diversity of food accepted by consumers. The knowledge of all mechanisms involved in food flavor changes during processing and storage period is compulsory. Ideally, food should have the same flavor from the moment of packaging to the moment it gets to the consumer. Unfortunately, food is exposed to chemical reactions starting with the moment it is taken out from the plant. Given the actual trend of big food industry producers to accede to the global food market, this involves longer durations on which the product passes until it arrives to the consumers, contributing in increasing risk of its quality damage. New innovative cost-effective technologies are in discussion, capable of satisfying the actual consumer's quality demands.

Foods were thermally treated for cooking purposes since many centuries ago, firstly to modify and preserve their organoleptic and nutritional characteristics. Anthropologists agree that thermal cooking of food had a great impact on phenotypical properties, intellectual development of mankind, and later to the society, depending on economy, culture, and geography. The transfer from home cooking to industrialized processes began in the late nineteenth century, firstly aiming food preservation and later, after the Second World War, focusing food safety and quality issues. Food engineering as a study discipline was first introduced in West European countries willing to offer a curriculum for actors in the food industry, including chemical engineering aspects, food microbiology, and biology. Nowadays, there is a difference between food technology and food engineering. The first treats the application of methods of food preservation and food processing, while food engineering is a more complex area focusing on a combination between food and applied sciences such as microbiology, physical sciences, chemistry, and engineering.

Thermal food processing is one of the most efficient methods of preservation. Beneficial effects include inactivation of food pathogens, natural toxins, and other unwanted changes, prolonging the shelf-life, adjusting the food digestibility, increasing the bioavailability of nutrients, and improving the functional properties. Still, thermal processing usually comes with unwanted changes in food composition such as decrease in nutrients, formation of toxic compounds, or other influencing the organoleptic properties. Continuously, heat treatment has to be improved according to the latest trends in food research, focusing on the increasing beneficial effects and counteracting the undesired effects.

Active packaging concept gains more interest due to its effectiveness in antimicrobial protection insurance, especially to pathogenic microorganisms. Still open for discussion, the subject refers to the influence of food-packaging polymers on food products after the incorporation of protective microorganisms. Actually, this is one of the biggest challenges when considering transposing these biomaterials to industrial scale use. The past decade offered a great opportunity in valorizing the results in active biofilms characterized by antimicrobial and antifungal activities in order to extend the food shelf-life and to reduce the addition of food additives. As so, biologically active polysaccharides such as kefiran or chitosan proved an incredible potential to food industry by their high antimicrobial activity against pathogenic and spoilage microorganisms. The positive impact relates also to better environmental practices by avoiding the conventional packaging materials such as plastics and reducing the food waste.

Advanced processing methods tend to preserve the characteristic properties of food including their organoleptic and nutritional qualities better when compared to the conventional food processing methods and, in some cases, are even more cost-effective. There is a clear rise in the populations suffering from food allergies, especially infants and children. Though this fact is widely attributed to the changing livelihood of population in both developed and developing nations and to the introduction of new food habits with advent of novel foods and new processing techniques, their complete role is still uncertain. As so, the return to ancient natural food recipes could be an alternative for reinterpretation of food in a more safety and healthier way. For example, cereals are traditionally processed into foods and beverages through fermentation worldwide. With lactic acid fermentation, as the most known ancient preserving method, studies found that valuable compounds are formed during the process. Additionally, lactic acid fermentation enhances food safety by reducing the toxic compounds and producing antimicrobial factors, which facilitate the inhibition and elimination of food pathogens. Many health-enhancing attributes are related to the consumption of lactic acid fermented products.

Nowadays, demands for nondairy substitutes with high acceptance and functionality, given by pleasant flavor and nutrient content (vitamins, minerals, antioxidants, prebiotics, and probiotics), respectively, are increasing. Cereal-based beverages have a great potential to fulfill this demand. As part of the worldwide traditional cuisine and culture, these products are often transposed to industrial re-interpretation as novel food products with various beneficial microorganisms (especially in the case of lactic acid fermented products) and functional properties designed for a balanced nutrition and life quality improvement. As so, through intensive research, focusing on this field and industrial production, these novel foods are taken beyond their national reputation. By applying novel trend technology, food products are designed for a balanced nutrition, easy and safe to store, prepare, and consume, and able to supply all the needed vitamins and nutrients.

The evolution of food industry offered many opportunities of using low-cost food additives and ingredients. Beverage industry is the most exposed to using caloric sweeteners, sucrose, high-fructose corn syrup, fruit juice concentrates, etc., which in latest studies proved their harmful health effects. Products such as soft drinks, carbonated soft drinks, fruit ades, fruit drinks, sports drinks, energy and vitamin water drinks, sweetened iced tea, cordial, squashes, and lemonade are attributed as the major sugar contributors in the USA for their consumption after each meal, as a supplementary source of calories. This intake severely contributes to obesity and obesity-related diseases in children and adults. As a result of novel food regulation, past years' research focused on the reduction of sugar in food and beverages along with avoiding the addition of other artificial sweeteners. Some prospects were made and proved the positive effects of a predicted reduction of sweeteners in food.

Wild edible plants as part of the cultural and genetic heritage are harvested and processed into value-added products especially in the Eastern Europe. They represent important sources of nutrients and health-promoting compounds, being consumed as remedies especially in rural and suburban areas. However, these products lack recognition as important contributors to human diet in developed areas. Current research promotes the using of wild edible plants in human diet, while several strategies present their importance for social, economic, and agro-ecological development. Given the growing interest for a natural alternative of products obtained by chemical synthesis, bioactive compounds found increasing application in food industry, cosmetics, pharmaceutical industry, and even in agriculture. Sea-buckthorn (Hippophaë rhamnoides), hawthorn (Crataegus monogyna), wild grown European blackberry (Rubus fruticosus L.), cornelian cherry (Cornus mas), blackthorn (Prunus spinosa), dog rose (Rosa canina), and bird cherry or hackberry (Prunus padus) are only few of the wild fruits with great potential for food industry as a valuable source of bioactive compounds, mainly phenolic compounds.

Novel bioactive compounds obtained by lab-scale extraction techniques such as high-pressure processing, pulsed electric field on berry fruits, grape pomace and food industry byproduct extraction, microwave-assisted extraction, ultrasonically assisted diffusion processes, supercritical fluid extraction, pressurized hot water extraction, gas-assisted extraction, and enzyme-assisted extraction were transposed to industrial application such as juice industry and oilseed processing industry. Extraction-dependent factors—temperature, infusion duration, tea, and water ratio—were also tested in tea brewing in order to deliver important product information to consumers.

Modeling of processes through mathematics is widely used in food industry. Recently, an experimentally validated multiscale modeling for coffee extraction was developed. Food loss is a major concern to all economic, environmental, and social aspects. In recent years, several modeling of waste treatments depending on food characteristics were proposed. In order to assess the stability of a food product during storage period, degradation kinetics modeling was elaborated using different indicators such as the ascorbic acid content and color intensity loss.

Food Preservation

Food preservation is one of the methods to protect food from unwanted microbial growth. After the food is produced, we store and protect by covering the rice and curry with lids to keep away flies and other insects. By this, we are protecting it from any infection caused by them. This is a short-term condition. Food preservation, on the other hand, is done to preserve food for a longer time.

Objectives of Food Preservation

Following are the important objectives of food preservation:

- To prevent microbial contamination.
- To kill pathogens.
- To minimise food spoilage and food poisoning.

Food Preservation Methods

Food preservation started long back in ancient times. Cooling, freezing, fermentation, sun-drying, etc., are few age-old food preservation techniques. With the advent of technology, modern methods of food preservation were developed. Chemicals and other natural substances were used for preservation. These substances are known as preservatives.

Methods of food preservation.

- Chemical Method: Salt and edible oils are two main preservatives which are used since ages to prevent microbial growth. This is why we add extra oil to pickles. Preservation by salt is known as salting. Salting helps to preserve fruits for a long term. Meats and fishes can also be preserved by salting. Other synthetic preservatives include vinegar, sodium benzoate, sodium metabisulphite, etc.
- Sugar: Sugar is another common preservative used in jams and jellies. Sugar is good moisture absorbent. By reducing moisture content, it restrains the microbial growth.
- Heat and Cold Methods: Boiling and refrigeration prevent around 70 percent of microbial growth. Boiling kills the microorganisms that cannot tolerate extreme temperatures. Thus, it helps in food preservation. Refrigerators have very low temperatures. Since microbes do not get optimum temperature they need for growth, their growth is inhibited. Pasteurization developed by Louis Pasteur is used until today to preserve milk.
- Smoking: Smoking prevents dehydration in fish and meat and thus prevents spoilage. The wood smoke contains a large number of anti-microbial compounds that slow the rancidification of animal fats.

- Canning: The food contents are sealed in an airtight container at high temperatures. Meat, fish, fruits are preserved by canning.
- Sterilization: This method is carried out to remove microbes from food. For e.g., milk sterilization at 100 °C kills the microbes.
- Dehydration: It is the process of removal of water from food. It is the simplest method and prevents food spoilage by removing water.
- Lyophilization: This is the process of freezing and dehydration of the frozen product under vacuum.
- Radiation: This method is also known as cold sterilization. The UV rays, X rays, gamma rays kill all the unwanted microbes present in food.

The principle behind all these methods is either to reduce the temperature conditions required for microbial growth or to evaporate the water content.

Food Processing

Food processing is the set of methods and techniques used to transform raw ingredients into food or food into other forms for consumption by humans or animals either in the home or by the food processing industry. Food processing typically takes clean, harvested crops or slaughtered and butchered animal products and uses these to produce attractive, marketable, and often long-life food products. Similar processes are used to produce animal feed. Extreme examples of food processing include the expert removal of toxic portions of the fugu fish or preparing space food for consumption under zero gravity.

Examples of some processed foods.

The benefits of food processing include the preservation, distribution, and marketing of food, protection from pathogenic microbes and toxic substances, year-round availability of many food items, and ease of preparation by the consumer. On the other hand, food processing can lower the nutritional value of foods, and processed foods may include additives (such as colorings, flavorings, and preservatives) that may have adverse health effects.

Food Processing Methods

Beer fermenting at a brewery.

Common food processing techniques include:

- Removal of unwanted outer layers, such as potato peeling or the skinning of peaches.
- Chopping or slicing, such as to produce diced carrots.
- Mincing and macerating.
- Liquefaction, such as to produce fruit juice.
- Fermentation, as in beer breweries.
- Emulsification.
- Cooking, by methods such as baking, boiling, broiling, frying, steaming, or grilling.
- Mixing.
- Addition of gas such as air entrainment for bread or gasification of soft drinks.
- Proofing.
- Spray drying.
- Pasteurization.
- Packaging.

Performance Parameters for Food Processing

When designing processes for the food industry, the following performance parameters may be taken into account:

- Hygiene, measured, for instance, by the number of microorganisms per ml of finished product.
- Energy consumption, measured, for instance, by "ton of steam per ton of sugar produced".

- Minimization of waste, measured, for instance, by the "percentage of peeling loss during the peeling of potatoes".
- Labor used, measured, for instance, by the "number of working hours per ton of finished product".
- Minimization of cleaning stops, measured, for instance, by the "number of hours between cleaning stops".

Benefits of Food Processing

More and more people live in the cities far away from where food is grown and produced. In many families, the adults are work from home and therefore there is little time for the preparation of food based on fresh ingredients. The food industry offers products that fulfill many different needs: from peeled potatoes that simply need to be boiled at home to fully prepared ready meals that can be heated up in the microwave oven in a few minutes.

Microwave oven.

Benefits of food processing include toxin removal, preservation, easing marketing and distribution tasks, and increasing food consistency. In addition, it increases seasonal availability of many foods, enables transportation of delicate perishable foods across long distances, and makes many kinds of foods safe to eat by de-activating spoilage and pathogenic micro-organisms. Modern supermarkets would not be feasible without modern food processing techniques, long voyages would not be possible, and military campaigns would be significantly more difficult and costly to execute.

Modern food processing also improves the quality of life for allergy sufferers, diabetics, and other people who cannot consume some common food elements. Food processing can also add extra nutrients such as vitamins. Processed foods are often less susceptible to early spoilage than fresh foods, and are better suited for long distance transportation from the source to the consumer. Fresh materials, such as fresh produce and raw meats, are more likely to harbor pathogenic microorganisms (for example, Salmonella) capable of causing serious illnesses.

Drawbacks of Food Processing

In general, fresh food that has not been processed other than by washing and simple kitchen preparation, may be expected to contain a higher proportion of naturally occurring vitamins, fiber and minerals than the equivalent product processed by the food industry. Vitamin C, for example, is destroyed by heat and therefore canned fruits have a lower content of vitamin C than fresh ones.

Tate & Lyle brand Corn Syrup being moved by tank car.

Food processing can lower the nutritional value of foods. Processed foods tend to include food additives, such as flavorings and texture enhancing agents, which may have little or no nutritive value, and some may be unhealthy. Some preservatives added or created during processing, such as nitrites or sulfites, may cause adverse health effects. Processed foods often have a higher ratio of calories to other essential nutrients than unprocessed foods, a phenomenon referred to as "empty calories." Most junk foods are processed, and fit this category. High quality and hygiene standards must be maintained to ensure consumer safety, and failure to maintain adequate standards can have serious health consequences. Processing food is a very costly process, thus increasing the prices of foods products.

Trends in Modern Food Processing

Health

- Reduction of fat content in final product, for example, by using baking instead of deep-frying in the production of potato chips.
- Maintaining the natural taste of the product, for example, by using less artificial sweetener.

Hygiene

The rigorous application of industry and government endorsed standards to minimize possible risk and hazards. In the U.S., the standard adopted is HACCP.

Efficiency

- Rising energy costs lead to increasing usage of energy-saving technologies, for example, frequency converters on electrical drives, heat insulation of factory buildings, and heated vessels, energy recovery systems.
- Factory automation systems (often Distributed control systems) reduce personnel costs and may lead to more stable production results.

Industries

Food processing industries and practices includes the following:

- Cannery,
- Industrial rendering,
- Meat packing plant,
- Slaughterhouse,
- Sugar industry,
- Vegetable packing plant.

Food Fortification

Food fortification has been implemented for a long period of time in industrialized countries to achieve the successful control of vitamin A and D deficiencies, several B vitamins (thiamine, riboflavin, and niacin), as well as iodine and iron. Salt irrigation was introduced in the early 1920s both in Switzerland and the United States and has progressively expanded worldwide since salt iodine is presently used in most countries. Since the early 1940s, the fortification of cereal products with thiamine, riboflavin, and niacin has become a common practice. Margarine was fortified with vitamin A in Denmark, while milk was fortified with vitamin D in the United States. Foods for young children were fortified with iron, thus substantially reducing the risk of iron-deficiency anemia in this age group. In recent years, folic acid fortification has become widespread in America, a strategy undertaken by Canada and around 20 Latin American countries. These approaches have proven to be effective in reducing the prevalence of many diseases due to deficiencies, such as goiter (iodine), xerophthalmia (vitamin A), rickets (vitamin D), and anemia (iron). Foods from around the world have begun to be fortified with calcium, iron, phosphorus, and vitamins (especially A, B, C, and D), depending on the chemical composition of the basic foods.

The need for food fortification is also presented by the WHO and FAO, among many other nationally recognized organizations. They have acknowledged the micronutrient deficiencies of more than 2 billion people, largely caused by a dietary deficiency of vitamins and minerals. The importance of these deficiencies for public health depends on their magnitude and spread, but the most vulnerable are pregnant women and young children, as fetal and child development, cognitive development, and resistance to infections are affected. Although the entire world population suffers from various nutritional deficiencies, people with low incomes are the most affected, particularly in developing countries, due to unsafe food consumption. Poverty, the lack of access to a variety of foods, and the lack of knowledge on appropriate dietary practices represent major drawbacks for socioeconomic development while also contributing to a vicious circle of underdevelopment. As such, long-term effects on health, learning, and productivity are significant, while they also generate a high level of social and public costs from reduced work capacity due to high rates of illness and disability.

It was also in less industrialized countries that fortification has become more and more appealing in recent years, so much that the planned programs have advanced more rapidly throughout the implementation phase than initially estimated. Bearing in mind the success of the relatively long program to fortify sugar with vitamin A in Central America, where the prevalence of vitamin A deficiency has been greatly reduced, there are similar initiatives in other regions of the world. Currently, the first attempt to fortify sugar in South Africa is taking place in Zambia, and if successful, it will be implemented elsewhere. Darnton-Hill and Nalubola identified at least 27 developing countries that could benefit from programs to fortify one or more foods.

According to researcher, the most efficient and accessible way of providing the population with vitamins and micronutrients is their additional fortification using these substances in consumer food products and daily foods, especially flour and bakery products. The fortification of foods should not diminish their nutritional qualities and quality, especially not to substantially alter the taste or assimilation of other nutrients contained therein, not to reduce shelf life or to change product harmlessness characteristics.

Biofortification

People require at least 22 mineral elements for their well-being. These micronutrients can be delivered through a proper diet. However, over 60% of the world's population is estimated to be iron-deficient(Fe), more than 30% are zinc-deficient (Zn), while 30% are iodine-deficient and 15% selenium-deficient (Se). In addition, calcium (Ca), magnesium (Mg), and copper (Cu) deficiencies are common in many developed and developing countries. These deficits influence many biochemical pathways directly or indirectly. This is due to the production of crops in areas with low mineral bioavailability and/or (sporadic) crop consumption with inherent tissue mineral concentrations, accompanied by the lack of fish or food products or low-micronutrient crops. At present, mineral malnutrition is considered to be one of the most serious global challenges that are active but avoidable. Mineral malnutrition can be addressed by dietary diversification, mineral supplements, food fortification and/or increased mineral concentration in edible crops (biofortification).

The term biofortification refers to the increase of the micronutrient concentration in the edible part of the plant and can be achieved both by using fertilizers and by stimulating the absorption of these minerals in the plant, while the use of micronutrient fertilizers was effective in increasing the micronutrient content in plants. Practically, the fortification of food is closely linked with the biofortification of plants (raw materials for food); the higher the nutritional value of the raw material, the smaller the need for fortification. Thus, from increasing the productivity of the cultivated varieties nowadays, the necessity of obtaining varieties with high nutritional value has become a very important task for agronomists, as almost half of the world population suffers from deficiencies in zinc, iron, and selenium.

In some countries, selenium wheat biofortification is achieved by using selenium-based fertilizers, as wheat is considered a major source of selenium for our daily diet. As such, selenium is considered to be an essential micronutrient for the human body, which can reduce the risk of degenerative diseases, including cancer. A study by Wu based on the consumption of common wheat biscuits biofortified with selenium through fertilization showed that although the levels of selenium in blood plasma increased, there were no significant changes of degenerative disease-specific biomarkers and of the health condition in general. Another method of biofortification that aimed at the development of wheat varieties enriched with some micronutrients was based on interactions between genotypes and environmental factors and led to good iron and zinc levels in humans.

Government agencies, as well as food policy makers, support food enrichment and fortification to different degrees, in order to lower the deficiency rate in populations on a large scale, this proving to be an efficient approach. It is estimated that 2 billion people worldwide suffer from micronutrient malnutrition. From a public health standpoint, it is estimated that micronutrient deficiencies account for approximately 7.3% of global diseases. The most important elements used for food fortification are:

Iron

The greatest part of the iron in the human body is found in erythrocytes as hemoglobin, where its main function is to carry oxygen from the lungs onto the tissues. Iron deficiency causes anemia, the most common and widespread nutritional disorder in the world and a public health problem in both industrialized and nonindustrialized countries. As a component of myoglobin, a protein that supplies oxygen to the muscles, iron supports metabolism. Additionally, iron is necessary for growth, development, normal cellular function, and the synthesis of hormones and the connective tissue. Dietetic iron has two main forms: heme and nonheme. Iron-fortified plants and foods contain only nonheme iron, while meat, seafood, and poultry contain both heme and nonheme iron.

People normally lose small amounts of iron through the urine, feces, gastrointestinal tract, and skin. Losses are higher in menstrual women due to blood loss. Hepcidin, a circulating peptide hormone, is the key element in regulating iron absorption and iron distribution in the body, including plasma. Hemoglobin concentrations lower than 13 g/dL in men and 12 g/dL in women indicate the incidence of iron-deficiency anemia (IDA).

Vitamin A

Vitamin A is the name of a group of retinoids soluble in fats, including retinol, retinal, and retinyl esters. Vitamin A represents an essential nutrient, thus necessary in small amounts for the normal functioning of the visual system in human beings, maintenance of cell growth function, epithelial cell integrity, immune function, and reproduction. Dietary vitamin A requirements are normally provided by a combination of preformed vitamin A (retinol), which is present in food of animal origin, and provitamin A carotenoids, which are derived from foods of plant origin, which have to be converted into retinol by tissues such as the intestinal mucosa and the liver, in order to be used by cells. Vitamin A deficiency is based on a frequent evaluation of serum or plasma retinol, and a deficit affects visual function, as vitamin A status indicators have traditionally been based on visual changes, especially night blindness and xerophthalmia. The World Health Organization (WHO) estimated that over 254 million preschool children worldwide have low serum retinol levels, and therefore these indicators may be considered clinically or subclinically deficiencies in vitamin A.

Two forms of vitamin A are available in the human diet: preformed vitamin A (retinol and its esterified form, retinyl ester) and provitamin A carotenoids. Preformed vitamin A is found in food from animal sources, including dairy products, fish, and meat (especially the liver). The most important provitamin A carotenoid is beta-carotene; other provitamin A carotenoids are alpha-carotene and beta-cryptoxanthin. The body transforms these pigments into vitamin A. Both provitamin A and preformed vitamin A should be metabolized intracellularly within the retina and retinoic acid, the active forms of vitamin A, to support the vital biological functions of the vitamin. Other carotenoids found in foods, such as lycopene, lutein, and zeaxanthin, are not turned into vitamin A.

Commonly, vitamin A deficiency occurs with a diet low in vitamin A sources (e.g., dairy products, eggs, fruits, and vegetables), poor nutritional status, and a high rate of infection, especially measles and diarrheal diseases. The best sources of vitamin A are foods of animal origin, especially liver, eggs, and dairy products containing vitamin A in the form of retinol, a form that can be easily used by the body. In fact, it is difficult for children to meet their vitamin A requirements if their diet is low in foods of animal origin and, similarly, if their diet is low in fat. Fruits and vegetables contain vitamin A in the form of carotenoids, the most important being β-carotene. In a mixed diet, the conversion rate of β-carotene to retinol is about 12:1. The conversion of other carotenoids to retinol is less effective, the corresponding conversion rate being 24:1. Different food preparation techniques, such as cooking, grinding, and adding oil, can improve the absorption of synthetic β-carotene food carotenoid in the oil, also widely used in vitamin A supplements. The latter has a conversion ratio of 2:1 to retinol, while the synthetic β-carotene forms, commonly used to fortify foods, have a conversion ratio of 6:1.

Analyses have shown that high-dose vitamin A supplements can reduce measles mortality by up to 50%. Another analysis found that improving vitamin A deficiencies, either by supplements or by fortification, decreased mortality by 23% in children aged 6 months to 5 years, irrespective of the cause. Another study on a group of preschoolers who consumed 27 g of vitamin A-fortified margarine for 6 months reported a reduction in the prevalence of low serum retinol concentrations from 26 to 10%. A plasma retinol concentration lower than 0.70 micromoles/L (or 20 micrograms/mcg/dL) reflects vitamin A deficiency in a population, whereas concentrations of 0.70–1.05 micromoles/L could be marginal in some people. RDA for men and women is 900 and 700 µg retinol activity equivalent/day. Tolerable upper intake levels for preformed vitamin A in adults are set at 3000 µg.

Iodine

Iodine is present in the body in small amounts, mainly in the thyroid gland, while its only confirmed role is in the synthesis of thyroid hormones. Iodine deficiency is a major public health problem for populations around the world but especially for young children and pregnant women. In some areas, it is a significant threat to national, social, and economic development, as the devastating result of iodine deficiency is mental retardation: currently one of the main causes of cognitive impairment that can be prevented. This is the primary reason behind the current global mission to eliminate iodine deficiency-related disorders. Iodine is an essential component of thyroid hormones, thyroxine, and triiodothyronine. Thyroid hormones regulate many important biochemical reactions, including protein synthesis and enzymatic activity, as critical determinants of metabolic activity. These are also necessary for an adequate development of the central nervous system and skeletal system in the fetus and infants.

Food iodine and iodine salt are present in several chemical forms, including sodium and potassium salts, inorganic iodine, iodate, and iodide, the reduced iodine form. Iodine rarely appears as an element but rather as a salt. For this reason, it is called iodide and not iodine. Iodine is rapidly and almost completely absorbed in the stomach and duodenum. Iodate is reduced in the gastrointestinal tract and absorbed as iodide. When the iodide enters the circulation, the thyroid gland concentrates it in quantities adequate for thyroid hormone synthesis, and most of the remaining amount is excreted in the urine. Healthy adults with sufficient iodine amounts exhibit approximately 15–20 mg of iodine, 70–80% of which are contained in the thyroid.

Average urinary iodine concentrations of 100–199 mcg/L in children and adults, 150–249 mcg/L in pregnant women, and > 100 mcg/L in nursing women indicate adequate iodine intake. Values lower than 100 mcg/L in children and adults who are not pregnant indicate insufficient iodine intake, although iodine deficiency is not classified as severe until the urinary iodine level is lower than 20 mcg/L. One of the best sources of iodine is marine algae, but its content is variable according to species. Other good sources include seafood, dairy products, cereal products, and eggs. Iodine is also present in human breast milk and in infant formulas.

Folate (Vitamin B9)

Folate (vitamin B9) is important in the synthesis and methylation of nucleotides that intervene in cell multiplication and tissue growth. Its role in protein synthesis and metabolism is closely interrelated to that of vitamin B12. The main sources of dietary folate are leafy vegetables, fruits, yeast, and liver. The total body content of folate is estimated at 10–30 mg; about half of this amount is stored in the liver, while the rest in the blood and body tissues. A serum folate concentration is commonly used to assess folic acid status, with a value greater than 3 ng/ml being adequate. The follicular erythrocyte concentration above 140 ng/ml indicates an adequate folate status, although some researchers have suggested that superior values are optimal for preventing neural tube defects.

Folate is naturally found in a wide variety of foods, including vegetables (especially leafy green vegetables), fruits and fruit juices, nuts, beans, peas, dairy products, poultry and meat, eggs, sea food, and cereal. Spinach, liver, yeast, asparagus, and brussels sprouts are among the foods with the highest levels of folic acid.

In January 1998, the US Food and Drug Administration (FDA) began to require producers to add folic acid to enriched bread, cereals, flour, corn, pasta, rice, and other cereal products. Since cereals are widely consumed in the United States, they have become very important contributors to folic acid intake for the American diet. The fortification program was designed to increase the folic acid intake in the United States by about 100 mcg/day but eventually succeeded an increase of about 190 mcg/day. In April 2016, the FDA approved the voluntary addition of folic acid to corn meal at levels compatible with other enriched cereal products. The Canadian Government has also called for the addition of folic acid to many cereals, including white flour, enriched pasta, and corn flour, as of November 1, 1998. Other countries, including Costa Rica, Chile, and South Africa, have also established compulsory programs for the folic acid fortification.

Vitamin B12

Vitamin B12 (cobalamin) is a cobalt-containing vitamin that is synthesized by microorganisms and exists in various chemical forms in food, particularly that of animal origin, such as milk, cheese and eggs, as well as artificially fortified foods. B12 deficiency can cause neurological damage, megaloblastic anemia, increase of homocysteine in plasma, and possibly impairment of the immune function. In infants and young children, it can cause serious developmental delays. Approximately

56% of an oral B12 dose of 1 mcg is absorbed, but absorption decreases drastically when the intrinsic factor capacity is exceeded (for 1–2 μm of vitamin B12).

Vitamin B12 is systematically evaluated through B12 serum or plasma levels. Approximate values below 170–250 pg/ml (120–180 picomoles/L) for adults indicate a vitamin B12 deficiency. Increased levels of methylmalonic acid (values>0.4 micromole/l) could be a more reliable indicator of vitamin B12 status, as they indicate a metabolic change that is highly specific to vitamin B12 deficiency. Vitamin B12 is naturally found in animal products, including fish, meat, poultry, eggs, milk, and dairy products. Generally, vitamin B12 is not present in plant foods, but fortified breakfast cereals are an available vitamin B12 source with high bioavailability for vegetarians. Some nutritional products from yeast also contain vitamin B12.

Other B Vitamins (Thiamine, Riboflavin, Niacin, and Vitamin B6)

In recent years, due to increased evidence that some B vitamins can prevent the occurrence of developmental disorders, as well as chronic degenerative and neoplastic diseases, special attention has been paid to their possibilities of employment. Moreover, vitamin B complexes are of utmost importance for energy metabolism. Specifically, thiamine (vitamin B1), riboflavin (vitamin B2), niacin (vitamin B3), and vitamin B6 are required for decarboxylation, transamination, acylation, oxidation, and reduction of substrates that are ultimately employed in energy consumption. One or several of these vitamins are also important for amino acid, fatty acids, cholesterol, steroid, and glucose synthesis. In fact, thiamine plays a critical role in energy metabolism and hence in cell growth, development, and function. Important sources can be found in wheat germs and yeast extracts, the edible organs of most animals, legumes, and green vegetables. The main sources of riboflavin are milk and dairy products, bread and bakery products, mixed meat-based foods, ready-to-eat cereals, and mixed grain-based foods; good sources of niacin are liver, meat and meat products, fish, peanuts, and whole grains. The richest sources of vitamin B6 include fish, beef liver and other organs, potatoes, and other starchy vegetables, and fruits (other than citrus fruits). Fortification of foods for nutrient recovery such as the vitamin B complex is effective in basic foods or spices, in wheat flour, fish sauce, and rice. RDAs are presented in table.

Vitamin C

Vitamin C is a redox system composed of ascorbic acid and dehydroascorbic acid, which acts as an electron donor. Its main metabolic function is to maintain collagen formation. Additionally, it functions as an important antioxidant. Vitamin C is widely available in food of plant and animal origin, but the best sources are fresh fruits and vegetables and the edible organs of animals. However, since vitamin C is unstable when exposed to an alkaline environment or to oxygen, light, and heat, losses can be substantial during storage and cooking. Foods fortified with vitamin C include milk and baby food, juices, jelly, and candies. Acute vitamin C deficiency leads to scurvy. Scurvy evolution time varies with vitamin C levels, but signs may occur within 1 month after the decrease or absence of vitamin C consumption (below 10 mg/day).

Vitamin D

Vitamin D deficiency can lead to musculoskeletal diseases such as rickets and osteomalacia, but vitamin D supplements and fortified foods can prevent extraskeletal disorders such as respiratory tract infections, asthma exacerbations, pregnancy complications, and premature death. Vitamin D has a unique metabolism as it is mainly synthesized in the skin under the influence of sunlight (i.e., ultraviolet radiation-B), while nutritional intake traditionally plays a relatively minor role. The recommended target concentrations range from ≥25 to ≥50 nmol/L (≥ 10–20 ng/ml), corresponding to a daily vitamin D dose of 10–20 μg (400–800 international units). Worldwide, vitamin D food fortification has already been introduced in the United States, Canada, India, and Finland with effective results. As such, foods recommended to be fortified with vitamin D include milk, dairy products, and margarine.

Very few natural foods contain vitamin D. Fatty fish (such as salmon, tuna, and mackerel) and fish liver oils are among the best sources. Small amounts of vitamin D are found in beef liver, cheese and egg yolks, and certain mushroom varieties. In the American diet, fortified foods provide most of the vitamin D intake. For example, almost all US milk is voluntarily fortified with 100 IU/cup. (In Canada, milk is fortified by law with 35–40 IU/100 ml, similar to margarine with ≥530 IU/100 g.) In the 1930s, a milk fortification scheme was implemented in the United States to fight rickets, considered to be a major public health problem. Other dairy products such as cheese are generally not fortified. Ready-to-eat breakfast snacks often contain vitamin D, similar to some brands of orange juice, yoghurt, margarine, and other foods. Both the United States and Canada require vitamin D-fortified formula for infants: 40–100 IU/100 kcal in the United States and 40–80 IU/100 kcal in Canada.

Calcium

Calcium is an essential element for the growth, activity, and maintenance of the human body. Bone health is a major public concern. Each year, approximately 9 million people worldwide suffer from fractures due to osteoporosis. Insufficient calcium and vitamin D absorption, an inadequate lifestyle, food choices, and genetics play an important role in the development of osteoporosis. Calcium absorption is controlled homeostatically by vitamin D regulation. Calcium deficiency can also lead to reduced blood clotting conditions, weak teeth, and some other symptoms. People with lactose intolerance have decreased calcium intake due to absent dietary intake of calcium-dairy products, while the availability of calcium from nondairy sources is affected by the presence of the phytic acid, oxalic acid, and fiber. Calcium is used for fortification in the form of calcium carbonate, calcium lysinate, and tricalcium phosphate in foods such as rice extrudates and noodles, tuna bone powder crackers, cookies, biscuits, yoghurt, etc.

The percentage of calcium absorbed depends on the total amount of elemental calcium consumed at one time; as the amount increases, the absorption percentage decreases. The absorption is highest in doses ≤500 mg. For example, someone who takes 1000 mg of calcium daily as supplements could divide the dose and take 500. The average dietary calcium intake for men over 1 year old is between 871 and 1266 mg/day, depending on the age group, while for females the interval is between 748 and 968 mg/day mg at two separate times during the day.

Selenium

Selenium (Se) is an essential micronutrient for both humans and animals. It forms an important component of glutathione peroxidase, a well-known antioxidant that counteracts cellular oxidative destruction. Furthermore, it plays an important role in catalyzing the production of the active thyroid hormone, and it is required to improve human immunity and sperm motility. Epidemiological studies have indicated that Se deficiency is positively correlated with the incidence of cancer. The average daily recommended dose for an adult is 60 μg/day. Selenium content in foods of plant origin varies with soil selenium content. The most prevalent selenium fortification is that of the cooking salt, which significantly reduced the prevalence of Keshan disease in China, but there are other foods that can be fortified with selenium, such as yoghurt.

Fibers

Constipation has a negative impact on the quality of life and is defined as a difficult evacuation accompanied by discomfort and pain, while long-term constipation entails serious health problems. This is due to a decrease in dietary fiber consumption, which is directly proportional to excessive food processing.

Food fibers can be classified as soluble and insoluble fibers. Both types of fiber have many health benefits including maintaining intestinal and overall health integrity, lowering blood cholesterol levels, controlling blood sugar levels, and providing a non-caloric volume that can help weight loss by replacing caloric food components such as fat. According to the Dietary Guidelines for Americans, dietary fiber is consumed by most adults indicating that fortification of fiber foods might bring health benefits. Numerous studies have shown that sausage fiber fortification (2–3 g/portion) can be achieved without any negative impact on sausage sensory quality. Alongside the increased fiber intake, products also come with other advantages such as fat replacement, increased water holding capacity, and improved oxidative stability when the fiber source is associated with phenolic antioxidants, bran bread, and fiber-containing yoghurt.

Proteins

Proteins belong to a category of biologically active compounds that are essential to life through their specific action. They are the building blocks for the formation of tissues in the human body (cell walls, muscles, blood, hair, internal organs such as the heart and the brain, etc.), hormones, enzymes and antibodies, and replacement of waste cells. The essential amino acids are those that the body cannot produce by itself, and thus, they must be secured from food. Nonessential amino acids are amino acids that can be synthesized by the human body from essential acids or by cleavage of proteins. The quality of plant proteins (vs. those from animal sources) has become a very debatable topic due to the consumption of plant products, which is increasingly promoted in the "world of nutrition." As the presence of essential amino acids in the body is of utmost importance, nutritionists recommend the right combination between the two sources (plant and animal). An advantage can be found in the addition of a vegetable matrix to meat preparation. This improves the mineral profile of the product and brings a supply of vitamins to the finished product, for example, mushroom sausages, meat products with soy protein isolate, and whey or wheat protein, but also the fortification of plant products with plant proteins, for example, the fortification of wheat flour with cottonseed, chickpeas, amaranth, quinoa, and lentils.

Fatty Acids

Essential fatty acids are important nutrients for the human body, especially for maintaining the health of the cardiovascular system. Several sources of information suggest that human beings have evolved on a diet that included a ratio of omega-6 to omega-3 fatty acids (EFA) of ~ 1, while western diets include a ratio of 15/1–16.7/1. Western diets exhibit omega-3 fatty acid deficiency, while they also include excessive amounts of omega-6 compared to the diet that human beings have evolved on and which set the foundation for their genetic patterns. Products fortified with fatty acids in balanced proportions include meat, oil, jelly, various sauces, etc.

Multiple Fortifications

Based on what is known about the prevalence of individual micronutrient deficiencies, there are generally multiple deficiencies of common micronutrients or in different population groups. Micronutrient deficiencies are more common in people on a diet low in foods of animal origin and therefore low in iron and bioavailable zinc intake, as well as calcium, retinol (vitamin A), vitamin B2 (riboflavin), vitamin B6, and vitamin B12 intakes. Commonly, poor diets, lacking fresh fruits and vegetables, yield deficiencies of vitamin C (ascorbic acid), β-carotene (provitamin A), and folic acid. Grain milling eliminates several nutrients, especially iron and zinc, various B complex vitamins (i.e., thiamine, riboflavin, and niacin), and folate. People on refined grain diets are at increased risk of deficiency in all these micronutrients. Thus, fortifications are cumulated, such as vitamin A and iron, the vitamin B complex, calcium, and vitamin D, if micronutrients are separated. If fortification is achieved with the addition of different raw or auxiliary materials, the finished product will benefit from multiple fortifications, for example, the fortification of bread or sausages with mushrooms—this product will additionally have the nutritional value for protein and fiber content, while bakery products with the addition of nut paste come with a high intake of lipids, fibers, and minerals. Most of the products fortified with unprocessed vegetable additions exhibit multiple fortifications.

Advanced Applied Technologies for the Fortification of Bakery Products

The lentil flour was obtained by grinding the lentils, thus obtaining a fine consistency; then flour mixtures were created according to figure.

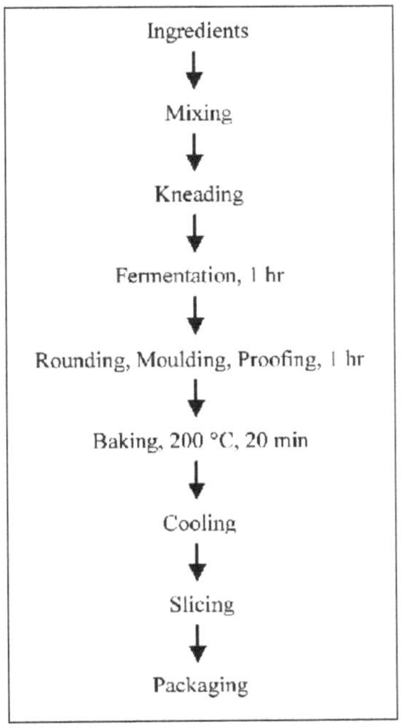

The technological chart for the production of bread fortified with lentil flour.

Several aspects need to be taken into account throughout the technological process of obtaining fortified bakery products. This process of obtaining different types of fortified bread, including lentil bread, can be achieved through a single-phase process, since the amount (10–30%) of lentil flour is always introduced into the process by uniform distribution in the flour mass before adding the wet components or through a two-phase process. The use of compressed yeast as suspension in water facilitates the even distribution of yeast cells in the dough, thereby contributing to the improvement of product quality. As such, preparing the compressed yeast involves performing several operations such as suspending, filtering the suspension, and activating the yeast. The dosing aims at obtaining the dough with the optimal rheological properties and the appropriate composition of the product. The dosing of the raw materials used for dough preparation is performed taking into account their physical characteristics.

During the kneading process, an amount of air is included in the dough, required to meet its rheological attributes. The physical processes that take place in the dough at its kneading are the mechanical action during the kneading that ensures the water penetration into the flour mass and the temperature increase of the dough. During the process of kneading itself, wet clumps of flour stick together under the influence of the mechanical kneading action, surface water penetrates deep, proteins become hydrated, the amount of bound water increases, and the dough increases its consistency and gradually acquires its elastic properties. The fermenting operation is aimed at obtaining risen dough that would produce suitable well-risen products with the right volume, as well as a porous and elastic core. Therefore, during this operation, various dough substances are accumulated in the dough, which impart the taste and aroma of bakery products. During dough fermentation, the yeast multiplication process continues, the predominant process being alcoholic fermentation, which releases carbon dioxide, ethyl alcohol, and a small amount of heat.

A well-risen product can only be obtained if the dough forms large amounts of gas. Ultimate leavening is the operation where the leavening of the dough reaches the maximum. The gases formed affect the gluten skeleton and thus the porosity of the dough by increasing or decreasing mesh size. The biochemical processes formed during fermentation are hydrolysis of starch and gluten amylolysis or proteolysis, respectively. The microbiological processes refer to the multiplication of yeasts and their fermentative activity, as well as that of lactic bacteria that make up dough microbiota. Colloidal processes are the continuation of the gluten-forming process and gluten peptization. Some of the gas present in the dough is lost during modeling, and as a result, the spongy structure of the dough is largely destroyed, the internal surface is reduced, and its specific weight increased.

A significant part of the carbon dioxide accumulated in the dough is eliminated during the division and modeling operations. Therefore, the dough must be subjected to a new fermentation for the restoration of its porous structure. As such, the products rise and the volume develops. The main goal of the final leavening of the dough is the rise by the accumulation of carbon dioxide that is formed during yeast alcoholic fermentation and conditions the volume and structure of the porosity of products.

Colloidal processes condition the dough to be transformed into the inner crumb and modify the water binding state so that the crumb, although exhibiting higher moisture than the initial humidity of the dough, appears dry due to thermal hydrophilic modifications. During coagulation, protein clotting and the gelatinization of starch occur. The protein coagulation process starts at temperatures slightly above 50° C and proceeds at a maximum speed of 60–70° C, thus accelerating to continue the heating of the dough. After coagulation, proteins become more easily attackable by enzymes, both digestive enzymes and the proteolytic ones in the dough. The gelatinization of the starch takes place in two stages: granule swelling and gelatinization itself. Swelling occurs due to the penetration of water groups within the starch granule that distances the protein chains. This is achieved by increasing the kinetic energy of water molecules by heating the dough. The gelatinization is accomplished by breaking the existing H2 bonds between the amylose and amylopectin chains and dispersing these chains between the molecules of water previously penetrated into the granule. Biochemical processes formed during baking are amylolysis and proteolysis. Microbial processes during baking include alcoholic fermentation and lactic fermentation.

Food Technology

Food Technology is a science branch that deals with the techniques involved in production, processing, preservation, packaging, labeling, quality management, and distribution of food products. The field also involves techniques and processes that are used to transform raw materials into food. Extensive research goes behind making food items edible as well as nutritious.

The complex food system is changing continuously. Processing of food has developed over the time making food easily accessible, more nutritious, safe and affordable. With the aid of various disciplines in food science and technology, raw materials are being converted to edible foods. Reduced diseases, improved safety and quality, variety, less cost, less wastage are some of the advancements in the field. But the growing need of food with specific attributes due to population hike is of concern. Biotechnology field has made progress to meet the increasing demand by improving crop production. Maintaining human health and wellness is of much importance. Newer technologies and detecting mechanisms can be anticipated in future with the evolving progress in nanotechnology. Minimizing risks and maximizing the benefits should be the main objective. Experts in Food science and technology should work hand in hand to get the maximum output. Investments in education and research are needed. The approach should be balanced and persistent to safeguard both nature and human health.

References

- The-growing-field-of-food-engineering-1345779: newengineer.com, Retrieved 21, July 2020
- Introductory-chapter-a-general-overview-on-latest-trends-in-food-engineering: intechopen.com, Retrieved 13, February 2020
- Food-preservation-methods-food-poisoning, biology: byjus.com, Retrieved 10, August 2020
- Food-fortification-through-innovative-technologies: intechopen.com, Retrieved 04, May 2020
- Food-technology, engineering-technology: aksuniversity.ac.in, Retrieved 25, January 2020

Chapter 2

Food Processing and Preservation Techniques

There are various principles which play a pivotal role in the preservation and processing of food. A few of them are ultrasonic processing, retort processing, heat preservation and processing, low temperature preservation and processing and Ohmic heating and inductive heating. This chapter discusses in detail these principles of food processing and preservation.

Food preservation includes preventing the growth of bacteria, fungi or other micro-organisms as well as retarding the oxidation of fats that cause rancidity thus promoting longer shelf life and reduced hazard from eating the food. Food processing is the transformation of raw ingredients, by physical or chemical means into food, or of food into other forms. Food processing is process of transformation of raw ingredients into food by the means of physical and chemical means. It is the process of producing raw food ingredients into marketable food products which can be easily prepared and used by the consumers. Example of Food Processing includes preparing space food for consumption under zero gravity. It is easy to keep processes food for long time, Canned and frozen fruits and vegetables, Foods fortified with nutrients such as fiber, vitamin D and omega-3 fatty acids. Safety is a major concern in food processing, because packaged food can be contaminated easily and due to increase in contamination quickly it can cause widespread illness. Food processing must be balanced with the food preservation also. Food preservation is to stop or slow down the spoilage of food, loss of quality and edibility of food for longer time. It involves preventing the growth of bacteria, fungi and microorganisms from. It includes drying, refrigeration, freezing, vacuum packing etc.

Ultrasonic Processing

Ultrasound is defined as sound waves having frequency that exceeds the hearing limit of the human ear (~20 kHz). Some animals utilize ultrasound for navigation (dolphins) or hunting (bats) using the information carried by back-scattering sound waves. Ultrasound is one of the emerging technologies that were developed to minimize processing, maximize quality and ensure the safety of food products. Ultrasound is applied to impart positive effects in food processing such as improvement in mass transfer, food preservation, assistance of thermal treatments and manipulation of texture and food analysis. Ultrasound (i.e., mechanical waves at a frequency above the threshold of human hearing) can be divided into three frequency ranges; power ultrasound (16–100 kHz), high frequency ultrasound (100 kHz–1 MHz) and diagnostic ultrasound (1–10 MHz).

In recent years, ultrasound technology has been used as an alternative processing option to conventional thermal approaches. Ultrasonication can pasteurize and preserve foods by inactivating many enzymes and microorganisms at mild temperature conditions, which can improve food quality in addition to guaranteeing stability and safety of foods. In addition, the changes to the physical properties of ultrasound, such as scattering and attenuation caused by food materials have been used in food quality assurance applications. Ultrasound is composed of sound waves with frequency beyond the limit of human hearing. By tuning frequency, ultrasound can be utilized in many industrial applications including food. Ultrasound techniques are relatively cheap, simple and energy saving, and thus became an emerging technology for probing and modifying food products.

Based on frequency range, the applications of ultrasound in food processing, analysis and quality control can be divided into low and high energy. Low energy (low power, low intensity) ultrasound has frequencies higher than 100 kHz at intensities below 1 W·cm2, which can be utilized for non-invasive analysis and monitoring of various food materials during processing and storage to ensure high quality and safety. Low power ultrasound has been used to non-destructively support genetic improvement programs for livestock and for evaluating the composition of raw and fermented meat products, fish and poultry. It is also used for the quality control of fresh vegetables and fruits in both pre- and postharvest, cheese during processing, commercial cooking oils, bread and cereal products, bulk and emulsified fat based food products, food gels, aerated and frozen foods. Other applications include the detection of honey adulteration and assessment of the aggregation state, size and type of protein. High energy (high power, high-intensity) ultrasound uses intensities higher than 1W·cm−2 at frequencies between 20 and 500 kHz, which are disruptive and induce effects on the physical, mechanical or chemical/biochemical properties of foods. These effects are promising in food processing, preservation and safety. This emerging technology has been used as alternative to conventional food processing operations for controlling microstructure and modifying textural characteristics of fat products (sonocrystallization), emulsification, defoaming, modifying the functional properties of different food proteins, inactivation or acceleration of enzymatic activity to enhance shelf life and quality of food products, microbial inactivation, freezing, thawing, freeze drying and concentration, drying and facilitating the extraction of various food and bioactive components.

Utilization of ultrasound in food technology for processing, preservation and extraction is such a system that has evolved to keep the wheel of development rolling. Ultrasound makes use of physical and chemical phenomena that are fundamentally different compared with those applied in conventional extraction, processing or preservation techniques. Ultrasound offers a net advantage in term of productivity, yield and selectivity, with better processing time, enhanced quality, reduced chemical and physical hazards, and is environmentally friendly. Microbial quality control is an important stage of the production chain for many food products in which biological contamination would have adverse effects on food quality. Acoustic streaming induced by ultrasound in a liquid is affected by the microbial activity, and thus, bacteria that alter the physicochemical parameters of a liquid (e.g. viscosity, elasticity) can be detected through the use of ultrasonic waves. Ultrasound refers to sound waves beyond the audible frequency range (in general, >20 kHz). When ultrasound passes through a liquid medium, the interaction between the ultrasonic waves, liquid and dissolved gas leads to an exciting phenomenon known as acoustic cavitation (AC). The main mechanism responsible for the ultrasonic deactivation effect is the physical forces generated by acoustic cavitation.

The asymmetric collapse of a cavitation bubble leads to a liquid jet rushing through the centre of the collapsing bubble. The speed of this microjet is a few hundred metres per second. Due to this high speed jet, pitting on solid surfaces has been observed. Microorganisms may have hydrophobic surfaces, which will promote the collapse of the cavitation bubbles on the surface leading to severe damage to the cell wall. Similarly, the microstreaming can lead to the erosion of cell walls, resulting in the inactivation of the microorganisms. High power ultrasound has only recently (<5 years) become an efficient tool for large scale commercial applications, such as emulsification, homogenization, extraction, crystallization, dewatering, low temperature pasteurization, degassing, defoaming, activation and inactivation of enzymes, particle size reduction and viscosity alteration. This can be attributed to improved equipment design and higher efficiencies of large scale continuous flow-through systems. Like most innovative food processing technologies, high power ultrasonics is not an off-the-shelf technology and therefore needs to be developed and scaled up for each application.

Properties of Ultrasonic Waves

Ultrasonic waves are high frequency waves that have a heating effect. They also have smaller wavelength than all the other waves and they form stationary wave pattern when passing through liquids. The ultrasonic waves have constant velocity in homogeneous medium. Many modes of vibration. High resolution used for flaw detection. Sonic beam and highly energetic.

- The ultrasonic waves cannot travel through vacuum.
- These waves travel with speed of sound in a given medium.
- Their velocity remains constant in homogeneous media.
- These waves can weld certain plastics, metals etc.
- These can produce vibrations in low viscosity liquids.
- The ultrasonic waves are reflected and refracted just like light waves.
- As ultrasonic waves cannot travel through vacuum, therefore if these waves travel through a non- homogeneous medium, then at each discontinuity like crack or change in density or presence of impurity etc., the amplitude and thus intensity of ultrasonic waves decreases by some amount. This decrease in intensity of ultrasonic waves as these travel through a medium is called Attenuation. The vacuum in the material causes strong reflection of ultrasonic waves while impurities or discontinuity cause the scattering of ultrasonic waves leading to net decrease in intensity. The attenuation is increased with increase in frequency of ultrasonic waves for a given medium. The intensity of ultrasonic waves decreases exponentially according to the relation.
- The speed of ultrasonic waves/acoustic waves is more in more dense media.
- Ultrasound travels through various media including gases, liquids and solids, but cannot travel through a vacuum. The speed of sound varies by the medium it travels through. Sound is likely to travel faster through solids, followed by liquids and gases. For example, the speed of sound in the air is about 340 meters per second (m/s). That in water is about 1530 m/s and that in iron as high as about 5,850 m/s. Another typical property of sound is that its energy is more likely to be lost in gases while it travels through liquids or solids more efficiently.

Table: Applications of ultrasound in food processing.

Applications	Conventional methods	Ultrasound principle	Advantages	Products
Cooking	Stove Fryer Water bath	Uniform heat transfer	Less time Improving heat transfer and organoleptic quality	Meat Vegetable
Freezing/ crystallization	Freezer Freezing by immersion, by contact	Uniform heat transfer	Less time Small crystals Improving diffusion Rapid temperature decreasing	Meat Vegetables Fruits Milk Products
Drying	Atomisation Hot gas stream Freezing Pulverisation	Uniform heat transfer	Less time Improving organoleptic quality Improving heat transfer	Dehydrated products (fruits, vegetables)
Pickling/ marinating	Brine	Increasing mass transfer	Less time Improving organoleptic quality Product stability	Vegetable Meat Fish Chees
Degassing	Mechanical treatment	Compression-rarefaction phenomenon	Less time Improving hygiene	Chocolate Fermented products (Beer)
Filtration	Filters (membranes semi-permeable)	Vibrations	Less time Improving filtration	Liquids (juices)
Demoulding	Greasing moulds Teflon moulds Silicon moulds	Vibrations	Less time Reducing products losses	Cooked products (cake)
Defoaming	Thermal treatment Chemical treatment Electrical treatment Mechanical treatment	Cavitation phenomenon	Less time Improving hygiene	Carbonated drinks Fermented products (Beer)
Emulsification	Mechanical treatment	Cavitation phenomenon	Less time Emulsion	Emulsions (Ketchup mayonnaise)
Oxidation	Contact with air	Cavitation phenomenon	Less time	Alcohols (wine, whisky)
Cutting	Knives	Cavitation phenomenon	Less time Reducing products losses Accurate and repetitive cutting	Fragile products (cake, cheese)

Ultrasound in Food Processing

Nowadays, the processed foods that are flourishing in supermarkets are modern processed foods and traditional foods, but their manufacturing, processing and packaging technologies have been advanced and rationalized to an incomparable extent. The principle aims of these technologies are to reduce the processing time, save energy and improve the shelf life and quality of food products. Thermal technologies (radio frequency and microwave heating), vacuum cooling technology, high pressure processing and pulsed electric field technology are those novel technologies who have potential for producing high-quality and safe food products but current limitations related with high investment costs, full control of variables associated with the process operation, lack of regulatory approval and importantly consumer acceptance have been delaying a wider implementation of these technologies at the industrial scale. In recent years, ultrasound (US) in the food industry has been the subject of research and development. There is a great interest in ultrasound due to the fact that industries can be provided with practical and reliable ultrasound equipment. Nowadays, its emergence as green novel technology has also attracted the attention to its role in the environment sustainability. Ultrasound applications are based on three different methods:

- Direct application to the product.
- Coupling with the device.
- Submergence in an ultrasonic bath.

Several processes such as freezing, cutting, drying, tempering, bleaching, sterilization, and extraction have been applied efficiently in the food industry. The advantages of using ultrasound for food processing, includes: more effective mixing and micro-mixing, faster energy and mass transfer, reduced thermal and concentration gradients, reduced temperature, selective extraction, reduced equipment size, faster response to process extraction control, faster start-up, increased production, and elimination of process steps. Food processes performed under the action of ultrasound are believed to be affected in part by cavitation phenomena and mass transfer enhancement.

Ultrasound has not only attracted considerable interest in the food industry due to its positive effects in processing, but more recently due to its promising effects in food preservation. Knorr shows successful reduction of E. coli in liquid whole egg using ultrasound. Generally, most micro-organisms showed greater sensitivity to ultrasound at increased temperatures over 50 °C. Elevated temperature weakens the bacterial membrane, which enhances the effect of cavitation due to the ultrasound. 'Ultrasonic pasteurization' at 50 °C has the potential of preserving the quality of many food products in terms of physicochemical properties, color and flavor compared to conventional pasteurization techniques at much higher temperatures.

Filtration

In the food industry, the separation of solids from liquids is an important procedure either for the production of solid-free liquid or to produce a solid isolated from its mother liquor. But the deposition of solid materials on the surface of filtration membrane is one of the main problems. The application of ultrasonic energy can increase the flux by breaking the concentration polarization and cake layer at the membrane surface without affecting the intrinsic permeability of membrane. The liquid jet serves as the basis for cleaning, and some other cavitational mechanisms lead to particle release from the blocked membrane. Ultrasonically assisted filtration (generally referred to as acoustic filtration) has been successfully employed to enhance the filtration of industrial wastewater that is generally considered difficult to process. Moreover, the optimized ultrasound intensity is very important to prevent the damage of filters. Ultrasound can also be applied to the production of fruit extracts and drinks. In the case of juice extraction from apple pulp, conventional belt vacuum filtration achieves a reduction in moisture content from an initial value of 85% to 50%, whereas electroacoustic technology achieved 38%. The use of ultrasound in combination with membrane filtration has also been investigated, with positive results. Here ultrasonic irradiation at low power levels was employed to aid the filtration of whey solutions.

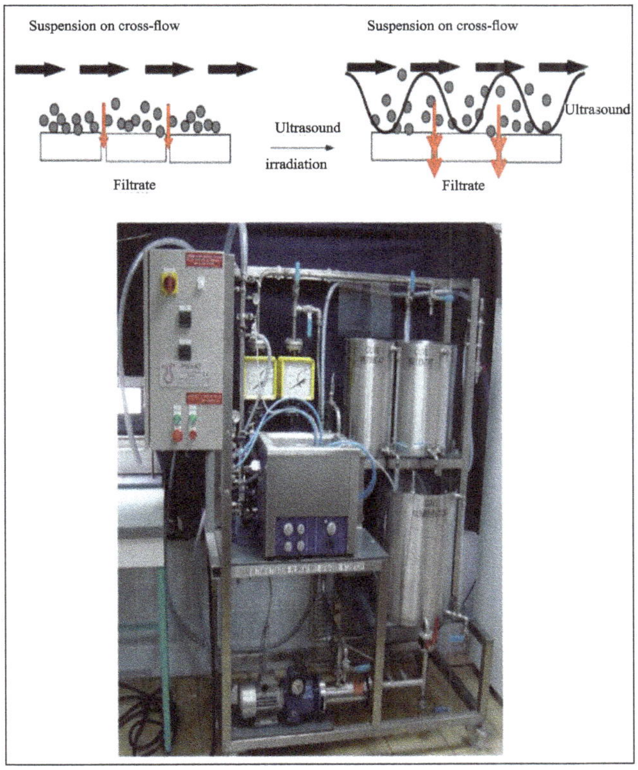

Enhancement of permeability using ultrasound.

Applications of Ultrasound

Meat Products

In the beef industry, ultrasound has been a fast, reproducible and reliable technology to enhance genetic improvement programs for livestock. Real time ultrasound has become widely used for estimation of the body and carcass chemical composition of growing lambs, sheep carcass and degree of muscle development in lamb and it has also been used for studying the composition of fish and poultry. Ultrasound used to determine the composition of chicken analogs and the solid fat content of chicken fat suggesting the advantages of LPU as a rapid and non-destructive method in food analysis. Ultrasonic velocity and temperature profiles were also used to study the composition of Atlantic mackerel (Scomber scombrus) tissues including fat content, solids and non-fat content. Ultrasound velocity measurements (UVM) have been used to determine the composition of raw meat mixtures. The quality of meat depends on the aroma, flavour, appearance, tenderness and juiciness. Consumer behavioural has shown that tenderness is most important palatability factor in determining meat quality. The traditional method used for meat tenderization is mechanical pounding, which makes poorer quality meat more palatable. Power ultrasound has also been found to be useful for this process. Ultrasound can act in two ways: by breaking the integrity of muscular cells or by enhancing enzymatic reactions, i.e. via a biochemical effect.

Sonicating beef muscle at $2 Wcm^2$ for 2 h at 40 kHz produced damage to the perimysal connective tissue, resulting in improved eating texture. Ultrasonic tenderization can be achieved with poultry meat, veal and beef. In the course of investigations into meat sterilization using heat and ultrasound, it was found that a side-effect was tenderization. It proved possible to reduce traditional heat treatment by 50% by using high-energy ultrasound. From these tests, it was concluded that an increased kill effect on micro-organisms could be achieved in various food test systems by using a combination of ultrasound and mild heating, a process known as thermosonication, or with pressure mano-sonication. Ultrasound has been used in the production of processed meats. In recombined meat products such as beef rolls, the pieces of muscle are held together by a protein gel, formed by the myofibrillar proteins released during processing. Tumbling the meat pieces and adding salt achieve this release. A sticky exudate is formed on the pieces, which binds them together when they are compressed. The binding strength, water holding capacity, product colour and yields were examined after treatment with either salt tumbling or sonication or both. Samples that received both salt treatment and sonication were superior in all qualities. Hence ultrasound can lead to improved physical properties of meat products, such as water-binding capacity, tenderness and cohesiveness.

Application of Ultrasound for Brining, Pickling and Marinating

Pickling and marinating are used for a wide variety of vegetables and meat products. Most current salt-brining or pickling-fermentation processes are subject to three main drawbacks:

- In brining, a very high sodium chloride content is required, which may require a 'desalting' process prior to shipping to reduce the sodium chloride content of the product.
- There is a potential lack of control in fermentation due to the occurrence of natural outside intrusion of 'wild' fermentation.
- Any soaking process can lead to enzymatic softening, structural damage and bloating.

All three of these side-effects are detrimental to rapid and efficient food preservation and so alternative technologies are of interest to food producers. Ultrasound allows the pickling time of products to be reduced considerably, particularly those foods with a crunchy texture. It also provides a method for manufacturing a pickle having a low level of sodium chloride compared with the pickles currently on the market. Hence there is no need to 'desalt', repack to reach the desired finished product salt level. Brining is a two way mass transfer process as the water migrates from the meat to the brine and solute from the brine to the meat. Pork loin slices were immersed in a saturated solution of NaCl at 21 °C for 45 min and different types of agitation of the solution and different levels of ultrasound intensity were applied during brining.

It was found that the water and NaCl contents of samples after treatment were higher in sonicated than non-sonicated samples. Moreover, ultrasound reduces the salting time, the formation of a crust and unwanted colouring of raw meat. The process also provides a product which is uniformly salted. In the cheese industry, the effect of ultrasound on mass transfer during cheese brining has been investigated. Many cheese varieties are salted by immersion in brine. Moreover, the influence of different process conditions, such as the use of agitation, brine concentration, sample: brine ratio and temperature, can be affected by acoustic energy. The rate of water removal and sodium chloride gain increased when ultrasound was applied in comparison with brining performed under static or dynamic conditions, suggesting that ultrasound improves both external and internal mass transfer.

Ultrasound extraction devices for liquid and solid products.

In the food industry, this Ultrasound technique can be used to degas carbonated beverages such as beer (defobbing) before bottling. In the processing of carbonated drinks, the purpose is to displace the air from the liquid surface in order to avoid organoleptic damage of the product by bacteria and oxygen. This process involves coupling a transducer to the outside of the bottle, leading to degassing. Compared with mechanical agitation, the ultrasonic method decreases the number of broken bottles and overflow of the beverage. The application of relatively lowintensity ultrasound during the fermentation of saké, beer and wine resulted is a reduction in the time required by 36–50%. Ultrasonically assisted degassing is particularly rapid in aqueous systems, but the removal of gas is much more difficult in very viscous liquids such as melted chocolate. Ultrasound technology has a wide range of current and future applications in the food industry. These industrial applications include texture, viscosity and concentrations measurements of many solid or fluid foods; composition determination of eggs, meats, fruits and vegetables, dairy and others products; thickness, flow level and temperature measurements for monitoring and control of several processes; and non-destructive inspection of whole fruits and vegetables, egg shells and food packages and also listed direct process improvements such as cleaning surfaces, enhancement of dewatering, drying and filtration, inactivation of microorganisms and enzymes, disruption of cells, degassing of liquids, acceleration of heat transfer and extraction processes, and enhancement of any process dependent upon diffusion of the ultrasonic extraction processes, biological cell.

Fruits and Vegetables

Plant foods including fruits and vegetables are highly attenuating materials due to the scattering of sound from voids and pores, which complicates the interpretation of ultrasound data and thereby unsuitable for evaluating their tissues. Mizrach, studied the application of ultrasound for the quality control of fresh vegetables and fruits in both pre- and postharvest applications and Mizrach explained the various physiological and physiochemical changes taking place during growth and maturation, and in the course of the harvest period, storage and shelflife, and how linking the results of ultrasound measurements and other physiochemical measurements, such as firmness, mealiness, dry weight percentage (DW), oil contents, total soluble solids (TSS), and acidity enables the indirect assessment of the proper harvesting time, storage period or shelf-life. An early study has shown that the amplitude of the ultrasound wave transmitted through fruit peels increased when the color changed from green to yellow indicating a good correlation between the ripeness and the acoustic attenuation. In other work, the maturity and sugar content of plum fruits determined by measuring ultrasound attenuation in the fruit tissue correlated well with the firmness of plums and that of tomato in other study. This proved the importance of using the attenuation parameter, which has also been used earlier for detecting defective potatoes. Ultrasound velocity measurements have also been used to determine the content of different sugar species in fruit juices and drinks. A non-contact ultrasonic system operating in either pulse–echo or through-transmission mode was used to measure the sugar content and viscosity of reconstituted orange juice. This system gave a good linear correlation with sugar contents in solution denoted by Brix, and an exponential correlation with viscosity. Another application of ultrasound velocity measurements is for the evaluation of oil composition, purity and quality. Sankarappa and coauthors measured the density and ultrasonic velocity at a frequency of 3 MHz in some refined and unrefined edible oils of coconut, castor, sunflower, safflower and groundnut, which allowed to estimate various physical parameters (e.g., specific volume, molar sound velocity, adiabatic compressibility, molar compressibility and intermolecular free length).

Low intensity ultrasound (50 or 100 kHz) has been used to assess the quality of various fruits and vegetables, including apples, tomatoes, carrots, avocadoes, mangoes and melons according to ripeness/texture/firmness. This was achieved by evaluating ultrasound parameters (velocity, attenuation) in relation to the physical characteristics (texture/firmness, total soluble solids, dry weight, and acidity) of the produce. Ultrasound has also been used as a pre-treatment prior to the drying of a range of vegetables. The treatment produced a reduction in subsequent conventional and freeze-drying times and also in rehydration properties. The peroxidase and polyphenol oxidase are the principal enzymes involved in the browning process of fresh-cut fruits and vegetables. Browning is important limiting factor in the consumers' acceptance. The study revealed that combined treatment of ultrasound and ascorbic acid had synergistic inhibitory effects on several enzymes related to enzymatic browning.

Ultra sound extraction method has also been developed for essential oil from aromatic plants such as peppermint leaves, or from other vegetal matrix such as garlic and citrus flowers. Increased yields of essential oil were found for peppermint leaves (up to 12%) and for artemisia when using Ultra sound, and increase by 2 to 3-fold of the main compounds of lavandula essential oil when comparing Ultra sound to conventional distillation. Moreover, Ultra sound extraction not only improved yields but as the method is fast and run at low temperature, the final product usually showed less thermal degradation than traditional method. Several studies have been run out on extraction of the main aroma compounds from spices. For example the vanillin was extracted from vanilla pods, carvone from caraway seeds, and safranal from Greek saffron. Yields of vanillin were comparables after one hour by UAE vs. 8 h in conventional extraction. Another publication showed that 80% of the pure vanillin was obtained after only 120 s of ultrasounds (ultrasonic probe) whilst it took 24 h with the conventional method to obtain 100%. For carvone extraction, the UAE method was compared to the Soxhlet. Unwanted fatty materials were extracted by Soxhlet extraction while the extract with Ultra sound extraction was of better quality and richer in carvone than in limonene. Wide varieties of fruits and vegetables have been studied by Ultra sound extraction because antioxidants are present in different amounts in different varieties of plants and these antioxidants come from different families.

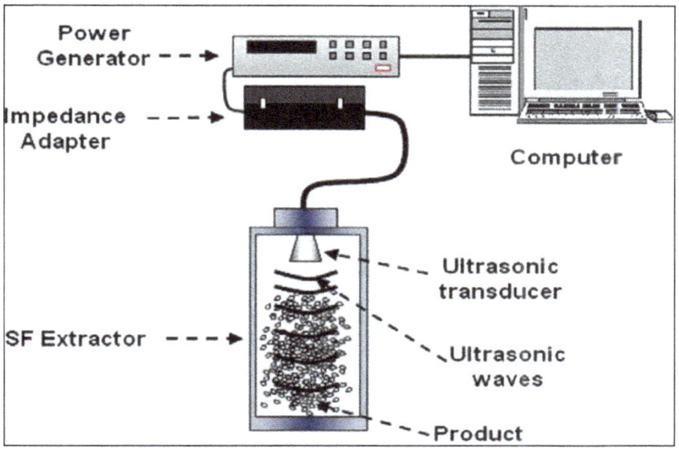

Scheme of the supercritical fluid extractor provided with an ultrasonic system to assist the extraction.

The effects of ultrasound could be related to the compressions and decompressions, the radiation pressure or the streaming. No clear evidence of cavitation was found under the conditions used for supercritical fluids. The high pressure needed to achieve supercritical conditions, above 72 bar, makes the appearance of cavitation bubbles difficult. The application of ultrasound affects the heat and mass transport processes. The effects linked to ultrasound include cavitation, compressions and expansions, microstirring, etc. and affect both the external and the internal heat and mass transfer resistance. The importance of each effect in the global influence of ultrasound on transport is different for the system considered: solid–liquid, solid–gas or solid-supercritical, since, for example, cavitation does not take place in a gas or supercritical medium. The process variables influence the magnitude of the ultrasound effects and it is necessary to establish the optimum value for each specific application. This offers new possibilities for food process innovation, ranging from energy savings to process yield or product quality. The use of ultrasound is opening up a field of activity in food processing. The main applications of ultrasound in food processes are linked to the effects it has on heat or mass transfer operations. Most of the ultrasonic applications reported in literature are found in liquid–liquid and liquid–solid systems due to the relative ease with which ultrasonic waves are transmitted in liquids. There is a wide offer of commercial equipment available on the market, including ultrasonic baths and different probe systems, which may be adapted for different operations. Thus, ultrasound has been applied in osmotic dehydration.

Fermentation

Several processes that take place in the presence of cells or enzymes are activated by ultrasonic waves. High intensity ultrasound can break cells or denature enzymes, however low intensity ultrasound can improve mass transfer of reagents and products through the boundary layer or through the cellular wall and membrane. Matsuura, Hirotsune, Nunokawa, Satoh, and Honda showed an increase in the fermentation rate of sake, beer and wine, when a relatively low intensity ultrasound was applied during the fermentation. The proposed mechanism is that the ultrasound (a great degassing tool) drives off CO_2 (produced during the fermentation) which normally inhibits the fermentation.

Future Trends

Research shows ultrasound can play an important role in food technology: processing, preservation and extraction. Although conventional cutting, emulsification and cleaning are often bottlenecks, lack of knowledge keeps industry from implementing ultrasound in their processes. A recent survey and market study of the possible future applications of new process technologies (like microwave, ultrasound) in the food industry has revealed that many companies are reluctant to apply these new technologies. The main reason is poor understanding of these new techniques by food professionals and the reason or weight of tradition. Commercial standard ultrasonic equipment is developing at great pace and no novel process for the application of ultrasound in industry is possible without ultrasonic equipment manufacturers willing to build new designs according to the requirements of customers.

Retort Processing of Food

A Retort is a pressure vessel used in the food manufacturing industry to "commercially sterilize" food after it has been placed into its container and the container has been hermetically sealed. More on commercial sterilization later. The method of sterilization is a thermal based process, or the application of high heat to the product. The high temperatures required in a sterilization process destroys pathogenic organisms that may be present in/on the container and/or food product, and is well above the boiling point of water at normal atmospheric pressure. So, the retort is by design a "pressure vessel". Thermal processing/sterilization of shelf stable, low acid foods (i.e.: defined as products with a finished equilibrium pH greater than 4.6 and a water activity greater than 0.85) is usually performed at temperatures at or above 250 °F or 121 °C, give or take ~10 °F/5.6 °C. The higher the temperature, the shorter the time the product needs to be exposed to heat.

Batch retorts are available in several configurations. First, they can be static or agitating style machines. A static retort is typically used for conductive (i.e. viscous) heating products where the product does not benefit from any kind of movement during the process (heat or cooling). Agitating style retorts are typically used for convective (i.e. "flowable liquid") type products that benefit from some product movement (i.e. "stirring") in the container during the process. Either from a process stand-point, to improve the rate of heat transfer into/out of the container, and/or from a product quality stand-point (i.e. to shorten the exposure time to heat). Agitating style retorts can utilize various methods of agitation depending on the orientation of the product container. Vertically oriented containers, such as cans or glass/plastic bottles, are typically agitated in a rotary fashion, either axially or end-over-end. With horizontally oriented containers, such as flexible pouches and semi-rigid bowls and/or trays (which contain product that benefits from agitation) the preferred method is horizontal reciprocation.

Batch retorts can employ various methods of process delivery. Some of which also utilize overpressure or counter-pressure to help protect the integrity of the container during the process (i.e. to keep the package from bursting as temperature and pressure builds inside the container during the process). Rigid containers, such as steel cans, can withstand large differences between the pressure inside and outside of the container, and therefore these types of containers typically do not require overpressure. They can be processed in a 100% saturated steam environment without the use of overpressure during the heating phases. On the other hand, more fragile flexible and semi-rigid containers cannot withstand high pressure differentials, so air is introduced into the retort to provide overpressure to maintain package integrity during the process. These types of containers require more sophisticated overpressure process delivery methods such as water spray, water cascade or water shower, water immersion or steam-air type systems. Because air is an insulator, a means of stirring or mixing the process media in the retort is required to avoid cold spots in the machine, thus ensuring good temperature distribution throughout the retort and product load. This mixing is accomplished by the different water flow methodologies, or via a fan in the case of steam-air retorts, and/or via mechanical rotation of the insert/drum in the case of agitating style machines.

Overpressure is also important in the cooling phases of a retort process because as cooling water is introduced into the retort it collapses the steam created in the heating step(s). Without sufficient introduction of air overpressure during cooling, the pressure in the retort can suddenly drop due to steam collapse thus creating a vacuum situation in the retort. If this happens the pressure differential between the outside environment and the temperature/pressure environment inside the container becomes too great thus causing the container to burst (otherwise known as "buckling"). Precise control of overpressure during the initial phases of cooling is important to avoid the above situation but ramping that pressure down in the latter stages of cooling is important as well to avoid crushing the container (or otherwise known as "paneling") as the temperature and pressure inside the container subsides. While the retort process inactivates or destroys bacterial pathogens, it does not destroy all microscopic spoilage organisms. Thermophiles are bacteria that can withstand temperatures well above typical retort temperatures. For this reason, the product must be cooled down to a temperature below that at which these organisms will reproduce, thus causing thermophilic spoilage.

Heat Preservation and Processing

The safety and storage life of many perishable foods can be enhanced by the use of high temperatures to inactivate undesirable disease and spoilage-causing microorganisms and to inactivate enzymes in food that can cause spoilage. Three categories of thermal preservation of foods are:

- Blanching.
- Pasteurization.
- Commercial Sterilization.

1. Blanching: Blanching is a form of thermal processing applied mainly to vegetables and some fruit by exposing them to heated or boiling water or even culinary steam for a short period of time. Blanching is a food processing operation designed to:

- Inactivate enzymes in plant tissues so that enzymatic degradation does not occur in the interval between packaging and thermal processing or during frozen storage or in the early stages of food dehydration and after reconstitution of dehydrated plant foods.
- Wilt vegetable products to enable packing of the products into containers so that proper fill weights can be achieved.
- Drive off inter- and intracellular oxygen and other gases from plant tissues so that containers are not deformed by excessively high internal pressures due to expanding gases within the container and to permit formation of a vacuum in the container after thermal processing.

2. Pasteurization: Pasteurization is a thermal process that involves using temperatures of at least 72 °C for 15 seconds (high temperature short time or HTST process), prior to packaging. Acid food products (pH < 4.6) are mainly pasteurized to inactivate spoilage causing microorganisms. Pathogenic microorganisms cannot grow and do not survive very well in acid foods such as citrus juices or apple juice (with the exception of Escherichia coli 0157:H7).

In low-acid food products such as milk, the basis for preservation by pasteurization is to inactivate pathogenic (disease causing) bacteria and viruses. Many spoilage-causing microorganisms can still survive typical pasteurization process conditions. For example, in milk, the proteolytic and lipolytic bacteria are more heat resistant and can survive the pasteurization process. This explains why the typical spoilage pattern of pasteurized milk reflects the proteolytic (protein degradation) and lipolytic (lipid degradation) action of the psychrotrophic, spoilage-causing bacteria. Because pasteurization does not kill all the psychrotrophic spoilage-causing bacteria in milk, pasteurized milk must be refrigerated to maintain shelf life quality. The durable life date on milk containers reflects the storage life that can be expected when milk is held at 4 °C or lower.

3. Commercial Sterilization (CS): The basis for preservation by Commercial Sterilization (CS) is to destroy both spoilage and disease causing microorganisms in low-acid and acid foods, thus rendering the food "commercially sterile".

Commercially sterile means the condition obtained in a food that has been processed by the application of heat, alone or in combination with other treatments, to render the food free from viable forms of microorganisms, including spores, capable of growing in the food at temperatures at which the food is designed normally to be held during distribution and storage". Therefore, commercially sterilization involves the destruction of spoilage-causing and disease-causing microorganisms. Commercially sterile foods may contain small numbers of thermophilic bacteria spores; however, the spores cannot germinate and produce actively growing cells at room temperature.

Canning

Canning involves heating the food with a minimum treatment of 121 °C moist heat for 15 minutes. The process usually involves pre-sealing the food in containers prior to heating. Canning can be traced back to the early 1800's. It is called the"botulinum cook". Today, if a can of food is being sterilized, each food particle must receive the heat treatment (e.g. 121 °C for 15 min). When food is placed in a can, the heat treatment will change since heat transfer to the food takes place at a slower rate. Depending on the size of the can, the time to achieve sterility could be several hours. Most commercially sterile products have a shelf life of 2 years or more.

Ultra-High Temperature Processing (UHT) and Aseptic Packaging

The combination of UHT and aspectic packaging is another form of commercial sterilization. It involves the application of "ultra-high temperature" (heat) to food before packaging, then filling the food into pre-sterilized containers in a sterile atmosphere. This process will render the food shelf stable or commercially sterile without the need for refrigeration. UHT- Aseptic packaging is a relatively new development whereby food can be heated to 140-150 °C very rapidly by direct injection of steam, held at that temperature for short period of time (e.g. 4-6 seconds) and then cooled, in a vacuum chamber to flash off the water added in the form of condensed steam. This is carried out as a continuous flow operation. The decrease in processing time due to the higher temperature, and the minimal come-up time and cool-down time leads to a higher quality product.

The UHT processed food is packaged aseptically into pre-sterilized containers. These are usually cartons made from laminated plastic, aluminum and paper, which are chemically sterilized with a combination of hydrogen peroxide and heat, and then filled in the same piece of equipment which is housed in a sterile environment. There are other forms of packaging that can also be used in aseptic UHT processing: plastic cans, flexible pouches, thermoformed plastic containers, bag-in-box, and bulk totes. UHT-aseptically packaged products have a shelf life of 6 months or more, without refrigeration. It depends on the type of packaging being used. For example, Tetra Pak cartons can eventually be more prone to perforations in the packaging layers, whereas the newer plastic bottles are more resistant to pin hole formation allowing them to have a longer shelf life.

Some examples of food products processed with UHT are:

- Liquid products: milk, juices, cream, yogurt, wine, salad dressings.
- Semi-liquid/solid products: baby foods; tomato products, fruits and vegetable juices, soups.

Contrary to popular opinion, UHT processed milk and juices do not contain added agents to provide the long storage life at ambient temperature in the laminated cartons. The products are preserved solely through the application of heat. It is critical that the sterilized products are transferred to packaging equipment under aseptic conditions, to avoid contamination after thermal processing. Please note that many products that are UHT treated are not aseptically packaged. This gives them the "advantage" of a longer shelf life at refrigeration temperatures compared to conventional pasteurized products. However, this does not produce a shelf-stable product at ambient temperatures due to the possibility of post-processing recontamination.

Selection of Heat Treatments

The intensity of the heat treatment employed for a particular food preservation application depends upon a number of factors. The main considerations in selecting the required temperature-time conditions for thermal processing are:

- What is the objective or purpose? (Blanching or pasteurization or commercial sterilization).
- Are there additional preservation steps? (Is it combined with other preservation method?)
- What are the physical, chemical properties of the food? (Type of food).
- What is the heat resistance of microorganisms in the food?

Here are some examples to illustrate these points:

Foods that will be consumed within a short period of time after processing can have storage life extended by a combination of pasteurization and refrigerated storage (used for pasteurized milk and for pasteurized, vacuum packaged, cured meats). Longer storage times at ambient temperatures in evacuated sealed containers requires the use of commercial sterilization. The time-temperature combination required for pasteurization and commercial sterilization is determined by the most heat-resistant disease-causing and spoilage-causing microorganisms in the particular food commodity.

For a particular food commodity, the type of thermal processing operation and the rate of heat penetration into the slowest heating portion of the food within a particular container are governed by the food's physical properties (solid vs. liquid, or solid particles suspended in a liquid) and chemical properties (pH, fat content, presence or absence of heat-inducible thickening agents, food components that have protective or antagonistic effects on the thermal resistance of microorganisms). It is imperative that thermal preservation processes be designed so that the slowest heating portion of the food commodity receives the specified time-temperature thermal treatment to minimize risks of illness and/or post-processing spoilage.

The thermal processes applied to foods are governed by the heat resistance of the microorganisms in the food. In low acid foods which are to be thermally processed and vacuum sealed within gas-tight containers, the microorganism of most concern is Clostridium botulinum. The habitat of Clostridium botulinum can be soil (agricultural and forest), water (fresh, brackish and marine) and mud (fresh water and salt water). As a consequence, all foods of agricultural and fisheries origin must be considered as being potentially contaminated with Clostridium botulinum spores.

Thermal preservation treatments of low acid foods that are to be packaged and stored under anaerobic conditions must be designed to ensure the destruction of any Clostridium botulinum spores that may be present in the food with the provision of a wide margin of safety. Actively growing vegetative Clostridium botulinum cells produce a very potent neurotoxin.

Destruction of C. Botulinum Spores

To determine the thermal resistance of heatresistant spores in foods, "Inoculated pack studies" are carried out using a non-pathogenic spore-forming bacterium, Clostridium sporogenes PA3679 (a putrefactive anaerobe). Since PA3679 spores are more heat resistant than those of Clostridium botulinum spores, a process designed to kill PA3679 spores will definitely kill Clostridium botulinum spores with a wide margin of safety.

Determining the Appropriate Temperature and Time Needed

Thermal Death Curves

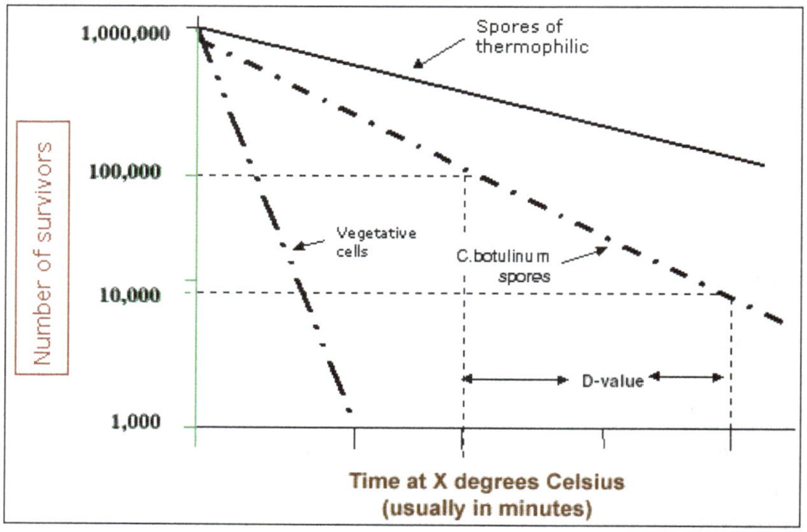

When microorganisms (bacteria, such as vegetative cells and spores, moulds, viruses and yeasts) are exposed to high temperatures capable of causing death of the organisms, one observes that the population is not killed instantaneously. We see that microbial death during thermal processing follows a logarithmic order. This means that bacteria are killed by heat at a rate that is nearly proportional to the number present in the system being heated.

The survivor curve or thermal death rate curve plotted in the figure depicts the logarithmic order of death. You will note that the time taken to traverse one logarithmic cycle represents the time, at a constant temperature, required to kill 90% of a microbial population. The exposure of microorganisms to a specific temperature for a period of time required to kill 90% of the population is defined as the decimal reduction time or D-value.

While the thermal death curve describes the rate of death of a particular microorganism under a specified set of conditions. A thermal death time curve can be constructed from a number of thermal death curves by exposing the microorganism to a variety of temperatures and determining the decimal reduction time at each temperature.

The thermal death time curve provides information about the time required to kill a particular microorganism in a particular food at a variety of temperatures. All of the time-temperature combinations along the plotted thermal death time curve represent the same killing power, with lower temperatures requiring longer time of exposure. You should also note that any point above the line (e.g., 100 minutes at 110 °C) will ensure that the microorganism is killed, while time-temperature combinations that fall below the plotted line (e.g., 100 minutes at 104 °C) represent conditions that will not kill all of the microorganisms present. The thermal death time curve that vegetative bacteria cells have a much lower heat resistance than spores.

Foods that are a combination of solids and liquid components heat up by a combination of convection and conduction heating. Note the location of the cold points in relation to conduction heating (in the geometric centre of a cylindrical container) and convection heating (one third of the way up the centre axis measured from the bottom of the cylindrical container). The cold point of a container of food must receive the required amount of thermal energy to ensure killing of Clostridium botulinum spores that may be present and to ensure a sufficient margin of safety. If a food formulation is changed such that the mechanism of heat transfer is altered from convection heating to conduction heating, and if the processing times are not altered to accommodate the change in mechanism of heat transfer, the food could be under processed and could pose a potential health hazard with respect to botulism.

Low Temperature Preservation and Processing

The preservation of food by refrigeration is based on a very general principle in physical chemistry: molecular mobility is depressed and consequently chemical reactions and biological processes are slowed down at low temperature. In contrast to heat treatment, low temperature practically does not destroy microorganisms or enzymes but merely depresses their activity. Therefore:

- Refrigeration retards spoilage but it cannot improve the initial quality of the product, hence the importance of assuring particularly high microbial quality.

- Unlike thermal sterilization, refrigeration is not a method of 'permanent preservation. Refrigerated and even frozen foods have a definite 'shelf life', the length of which depends on the storage temperature.
- The preserving action of cold exists only as long as low temperature is maintained, hence the importance of maintaining a reliable 'cold chain' all along the commercial life of the product.
- Refrigeration must often be combined with other preservation processes (the 'hurdle' principle).

Food preservation at low temperature comprises two distinct processes: chilling and freezing. Chilling is the application of temperatures in the range of 0 °C to 8 °C, i.e. above the freezing point of the food, while freezing uses temperatures well below the freezing point, conventionally below 18 °C. The difference between the two processes goes beyond the difference in temperature. The stronger preserving action of freezing is due not only to the lower temperature but also and mainly to the depression of water activity as a result of conversion of part of the water to ice. The use of refrigeration in the food industry is not limited to preservation. Refrigeration is applied for a number of other purposes such as hardening (butter, fats), freeze concentration, freeze drying, air conditioning including air dehumidification and cryo-milling.

Effect of Temperature on Food Spoilage

The relationship between the temperature and the rate of chemical reactions is described by the Arrhenius equation:

$$k = A\,e^{-\frac{E}{RT}} \Rightarrow \ln k = \ln A - \frac{E}{RT}$$

Where,

k = Rate constant of the chemical reaction.

E = Energy of activation, J/mol.

T = Absolute temperature, K.

R = Gas constant.

A = Frequency factor, almost independent of temperature therefore constant for a given reaction.

The kinetics of chemical spoilage reactions such as non-enzymatic browning and loss of some vitamins during storage is found to fit the Arrhenius model quite closely within a fairly wide temperature range. The Arrhenius model is therefore widely used for the prediction of chemical spoilage during storage, following non-monotonous time temperature profiles. Attention should be paid to the possibility of discontinuity in the Arrhenius model due to phase transition phenomena.

Effect of Low Temperature on Enzymatic Spoilage

The relationship between enzymatic activity and temperature follows the well-known bell-shaped curve, with maximum activity corresponding to the optimal temperature characteristic to each enzyme. On first sight, this behavior may seem to be in contradiction with the Arrhenius model. In reality, however, the bell-shaped behavior is the result of two simultaneous and contrary processes, both depressed by low temperature. The first is the enzyme-catalyzed reaction itself. The second is the thermal inactivation of the enzyme.

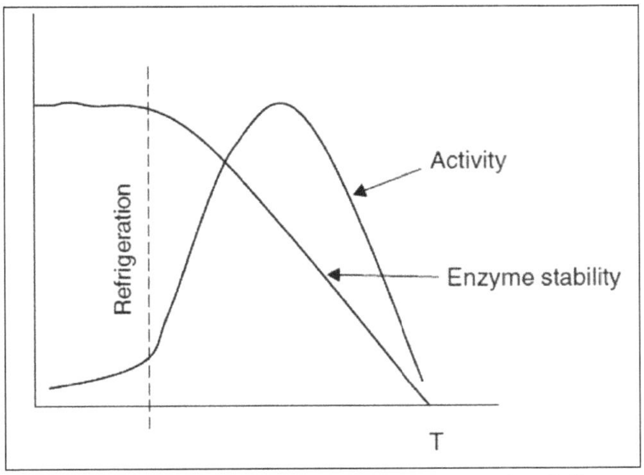

Effect of temperature on enzyme activity and stability.

Enzyme activity is strongly slowed down but not totally eliminated by refrigeration, even below freezing temperature. This is the reason for the need to inactivate the enzymes by blanching, particularly in frozen vegetables. Enzymatic activity in chilled and frozen foods is of considerable technological significance. Such activity may be desirable, as in the case of the aging of chilled meat or in the development of flavor in many kinds of cheeses but it can also be a source of deterioration such as the activity of proteases in fish or lipases in meat etc.

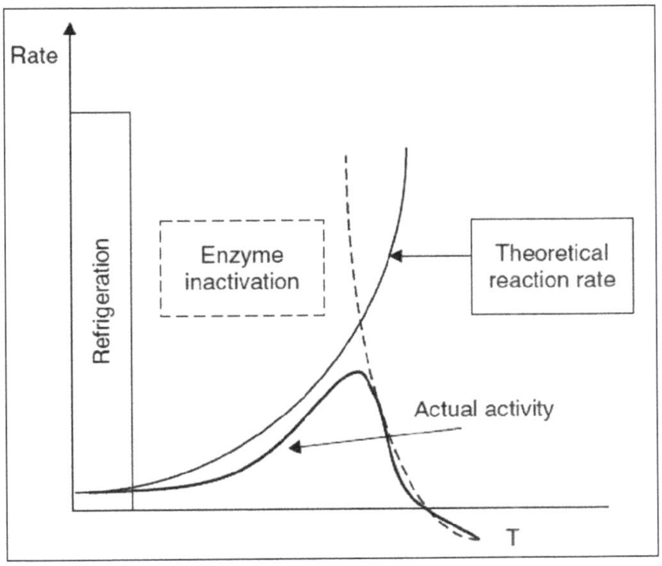

Bell shaped activity curve.

Effect of Low Temperature on Microorganisms

With respect to the effect of temperature on their activity, microorganisms are grouped into four categories: thermophiles, mesophiles, psychrotropes and psychrophiles. Typical growth temperature ranges corresponding to the four groups are given in Table. The four groups differ not only in their temperature requirement for growth but also in their rate of growth. Figure shows qualitatively the dependence of growth rate (generations per hour) and temperature for thermophiles, mesophiles and psychrophiles. Psychrophilic bacteria grow much more slowly, even at their optimal temperature. In refrigerated foods, psychrotropic and psycrophilic microorganism are obviously the main reason for concern.

Table: Bacteria, grouped by their growth temperature.

Group	Growth temperature (°C)		
	Minimum	Optimum	Maximum
Thermophiles	34 to 45	55 to 75	60 to 90
Mesophiles	5 to 10	30 to 45	35 to 47
Psychrotropes	– 5 to 5	20 to 30	30 to 35
Psychrophiles	– 5 to 5	12 to 15	15 to 20

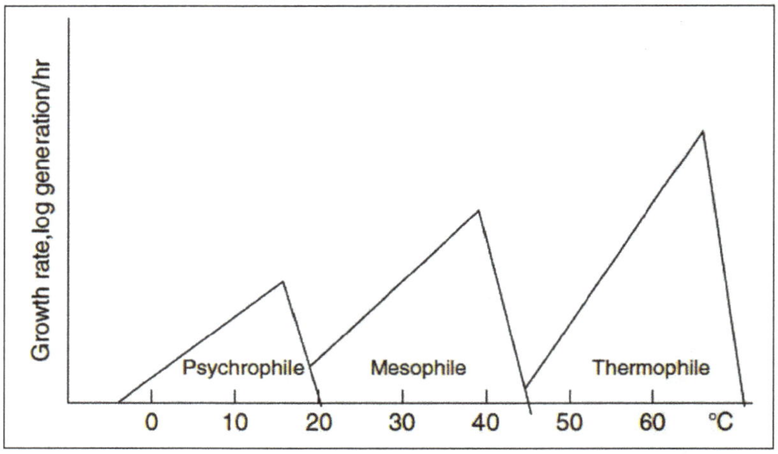

Classification of microorganisms by their response to temperature.

The relationship between temperature and shelf life is evident from the graph in Figure. It can be seen that, at lower temperature:

- The induction period (lag phase) is longer.
- Growth rate in the logarithmic phase is slower.
- As a result of the above, the bacterial count after a given storage time is considerably lower.

The microbiological quality of a refrigerated food depends therefore on the time temperature profile of the life of the product. Obviously, additional factors such as pH and water activity play an important role in the microbial processes in foods. Predictive microbiology is a new science, trying to develop tools for the prediction of the microbiological quality of foods, as a function of storage conditions.

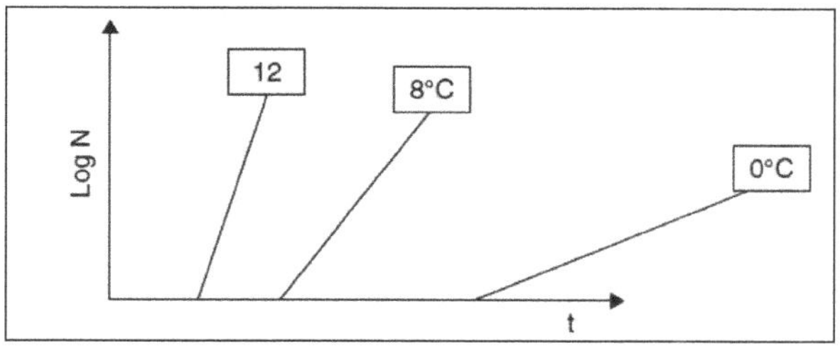

Schematic representation of the effect of storage temperature on the microbial load of a food.

Effect of Low Temperature on Biologically Active (Respiring) Tissue

By 'active tissues', we refer to foods such as fruits and vegetables after harvest or meat after dressing, in view of their biochemical activity. The principal biochemical post-harvest process that occurs in fruits and vegetables is respiration, whereby sugars are 'burnt ', oxygen is consumed and carbon dioxide is evolved. The rate of respiration is usually determined by measuring the rate of oxygen consumption or carbon dioxide release. Respiration is the most important (but not the only) cause of deterioration of fruits and vegetables during storage. It is usually said that the shelf life of fresh produce is inversely related to the rate of respiration. The rate of respiration is closely related to the temperature. The rate of respiration increases by 2 to 4 fold for every 10 °C increase in the temperature within the range of usual storage temperatures. However, too low storage temperatures may cause a condition known as 'chill injury' in certain commodities and must be avoided. Fruits and vegetables differ in the intensity of their post-harvest respiration rate. Following is a rough classification of some fresh produce with respect to their respiration rate during storage:

- High respiration rate: Avocado, asparagus, cauliflower, berries.
- Medium respiration rate: Banana, apricot, plums, carrot, cabbage, tomatoes.
- Low respiration rate: Citrus, apples, grapes, potatoes.

Many crops undergo a process of post-harvest ripening during storage. These are called 'climacteric' species, because their rate of respiration increases to a maximum in the process. Ethylene is produced and acts as a ripening hormone. Apples, bananas and avocado are climacteric, while citrus and grapes are non-climacteric. It is possible to control the rate of respiration of fruits, vegetables and cut flowers with the help of refrigeration alone or through the combination of refrigeration and 'controlled atmosphere' storage. 'Controlled atmosphere' consists of long-range refrigerated storage in closed chambers where the atmosphere has been artificially modified so as to depress the rate of respiration to the desired level. The optimum composition of the atmosphere depends on the commodity and often on the particular variety of the commodity. Table shows typical optimal storage conditions for selected fruits and vegetables.

Table: Optimal storage conditions for some fruits and vegetables.

Fruits in controlled atmosphere			
Produce	Temperature	Atmosphere	
		% Oxygen	%CO_2
Pineapple	12 – 15	5	10

Avocado	12 – 15	2–5	3–10
Pomelo	10 – 15	3–10	5–10
Lemon	10 – 15	5	0–10
Mango	10 – 15	5	5
Papaya	10 – 15	5	10
Melon (cantaloupe)	5 – 10	3–5	10–15
Vegetables in ordinary cold storage			
Produce	Temperature (°C)	Relative humidity (%)	
Artichoke	0–2	90–95	
Asparagus	0–2	95–100	
Broccoli	0–2	95–100	
Carrot	0–2	98–100	
Eggplant	10–14	90–95	
Onion	0–2	65–75	
Potato	8–12	90–95	

Respiration is an exothermic process. The heat released by respiring commodities during refrigerated storage and transportation must be taken into account in calculating the refrigeration load required. Table gives the approximate rate of heat evolution during storage of selected commodities. Low temperature alone is not always sufficient for extending the shelf life of fruits and vegetables. Another important condition that must be controlled in post-harvest storage is relative humidity. Loss of water is often the cause of texture deterioration, wilting, shriveling etc. Water loss may be minimized by storing at high relative humidity, but humidity levels that can promote growth of mold must be avoided.

Table: Approximate rate of heat evolution during refrigerated storage of selected commodities, according to storage temperature.

Commodity	Heat evolution rate, Watts per ton, when stored at:			
	0 °C	5 °C	10 °C	15 °C
Apple	10–12	15–21	41–61	41–92
Cabbage	12–40	28–63	36–86	66–169
Carrot	46	58	93	117
Sweet corn	125	230	331	482
Green peas	90–138	163–226	–	529–599
Orange	9	14–19	35–40	38–67
Strawberry	36–52	48–98	145–280	210–275

The Effect of Low Temperature on Physical Properties

Numerous changes in physical properties caused by exposure to low temperatures may have significant effects on food quality. Some of these changes are:

- Increase in viscosity.
- Decrease in solubility, resulting in crystallization, precipitation, cloudiness (e.g.in beer).

- Hardening, transition to rubbery and glassy state in carbohydrate systems.
- Hardening of fats.
- Decomposition of colloidal systems such as emulsions and gels.

Freezing

Freezing is one of the most widespread industrial methods of food preservation today. The transition from chilling to freezing is not merely a continuous change that can be explained on grounds of the lower temperature alone. On the contrary, freezing represents a point of sharp discontinuity in the relationship between temperature and the stability and sensory properties of foods.

- The exceptional efficiency of freezing as a method of food preservation is, to a large extent, due to the depression of water activity. Indeed, when food is frozen, water separates as ice crystals and the remaining non-frozen portion becomes more concentrated in solutes. This 'freeze concentration' effect results in the depression of water activity. In this respect, freezing can be compared to concentration and drying.
- On the other hand, the same phenomenon of 'freeze concentration' may accelerate reactions, inducing irreversible changes such as protein denaturation, accelerated oxidation of lipids and destruction of the colloidal structure (gels, emulsions) of the food.
- The rate of freezing has an important effect on the quality of frozen foods. Physical changes, such as the formation of large ice crystals with sharp edges, expansion, disruption of the osmotic equilibrium between the cells and their surroundings, may induce irreversible damage to the texture of vegetables, fruits and muscle foods. It has been established that such damage is minimized in the case of quick freezing.

Effect of Freezing and Frozen Storage on Product Quality

For a very large number of food products, freezing represents the best preservation method with respect to food quality. The nutritional value, the flavor and color of foods are affected very slightly, if at all, by the process of freezing itself. The main quality factor that may be adversely affected by freezing is the texture. On the other hand, unless appropriate measures are taken, the deleterious effect of long-term frozen storage and of thawing on every aspect of product quality may be significant.

Effect of Freezing on Texture

In a vegetal or animal tissue, the cells are surrounded by a medium known as the extracellular fluid. The extracellular fluid is less concentrated than the protoplasm inside the cells. The concentration difference results in a difference in osmotic pressure which is compensated by the tension of the cell wall. This phenomenon, known as turgor, is the reason for solid appearance of meat and the crispiness of fruits and vegetables. When heat is removed from the food in the course of freezing, the extracellular fluid, being less concentrated, starts to freeze first. Its concentration rises and the osmotic balance is disrupted. Fluid flows from the cell to the extracellular space. Turgor is lost and the tissue is softened. When the food is thawed, the liquid that was lost to the extracellular space is not reabsorbed into the cell, but is released as free juice in the case of fruits or 'drip' in the case of meat.

It is generally believed that freezing damage to the texture of cellular foods is greatly reduced by accelerating the rate of freezing. This is explained by the fact that the condition of osmotic imbalance, created at the onset of freezing, disappears when the entire mass is frozen. Furthermore, rapid freezing results in the formation of smaller ice crystals, presumably less harmful to the texture of cellular systems. Another probable reason for the deterioration of the texture is the volumetric expansion caused by freezing. The specific volume of ice is 9% higher than that of pure water. Because cellular tissues are not homogeneous with respect to water content, parts of the tissue expand more than the others. This creates mechanical stress that may result in cracks. Obviously, this effect is particularly strong in foods with high water content such as cucumbers, lettuce and tomatoes. This kind of texture damage is partially prevented by adding solutes. Adding sugar to fruits and berries before freezing was a widespread practice before the development of ultra-rapid freezing methods.

The rate of freezing affects the size of ice crystals. Slow freezing produces large crystals. It has been suggested that large crystals with sharp edges may break cell walls and contribute to texture deterioration in cellular foods subjected to slow freezing. There is no controversy as to the fact that slow freezing results in higher percentage of drip in meat and fish. It is also accepted that rapid freezing causes less damage to the texture of particularly fragile fruits. On the other hand, the general applicability of the theory stating the superiority of rapid freezing seems to be questionable. Nevertheless, quick freezing continues to be the practical objective of food freezing process design.

Effect of Frozen Storage on Food Quality

Frozen storage, even at fairly low temperature, does not mean the absence of deteriorative processes. On the contrary, frozen foods may undergo profound quality changes during frozen storage. While the rate of reactions is generally (but not always) slower in frozen foods, the expected shelf life, and therefore the time available for the reactions to take place, is long. Some of the frequent types of deterioration in frozen foods are protein denaturation resulting in toughening of muscle foods, protein–lipid interaction, lipid oxidation and oxidative changes in general (e.g. loss of some vitamins and pigments).

Extensive studies on the effect of frozen storage on product quality were undertaken in the 1960s by researchers. A large number of commodities were tested for changes in chemical composition and sensory characteristics. A concept known as 'time temperature tolerance' (TTT) was developed. The studies showed a linear relationship between the storage temperature and the logarithm of storage time for equal reduction in quality (loss of a certain vitamin, loss of color or loss of organoleptic score). The logical conclusion of these studies could be that lower storage temperature always results in higher quality. This is not always the case, however. Reactions may be accelerated by the 'freeze concentration' effect more than they are attenuated by the lower temperature. In this case, the rate of deterioration (e.g. lipid oxidation) may increase as the storage temperature is lowered, passing through a maximum and then diminishing at very low temperatures. Mass transfer phenomena during frozen storage (oxygen transfer, loss of moisture) may be a major cause of quality loss. The quality of packaging is, therefore, particularly important in frozen foods. The PPP (product–process–package) approach consists of paying attention of all three factors in evaluating and predicting the effect of frozen storage on product quality.

Another type of change in frozen foods during storage is the process of recrystallization. Smaller crystals are more soluble than large ones. Equally, small ice crystals have a lower melting point then large ones. Consequently, if the storage temperature undergoes fluctuations, small ice crystals may melt and then solidify on the larger crystals. This may be the reason why foods frozen rapidly and those frozen slowly sometimes show similar ice crystal size distribution after storage. Recrystallization is particularly objectionable in ice cream, where conversion of small ice crystals to large ones results in loss of the smooth, creamy texture. The remedy is, of course, avoiding temperature fluctuation during storage as much as possible.

Freezing Methods

There are three major methods of freezing foods:

- Air Freezing:
  - Still-air 'sharp' freezer,
  - Air blast freezer,
  - Fluidized-bed freezer (IQF).
- Indirect Contact Freezing:
  - Single plate,
  - Double plate,
  - Pressure plate,
  - Slush freezer kill.
- Immersion and Cryogenic Freezing:
  - Heat exchange fluid,
  - Compressed gas,
  - Refrigerant spray.

Air Blast Freezing

Air freezing is the oldest and most common type of freezing used. The freezer section of household refrigerator and the deep freezer are examples of still air freezers or low air velocity systems.

Air blast freezing is a moderately fast freeze because of vigorous circulation of cold air. The product is placed on trays or mesh belts and passed slowly through an insulated tunnel. In different systems the temperature may range from -18 °C to - 34 °C, with an air velocity of 100-3500 lineal feet per minute, with a counter current air flow. Air blast freezers operate at lower temperatures than still air freezers and rely on movement of the cold air at high velocity over the food in order to achieve rapid removal of heat and to maximize the freezing rate.

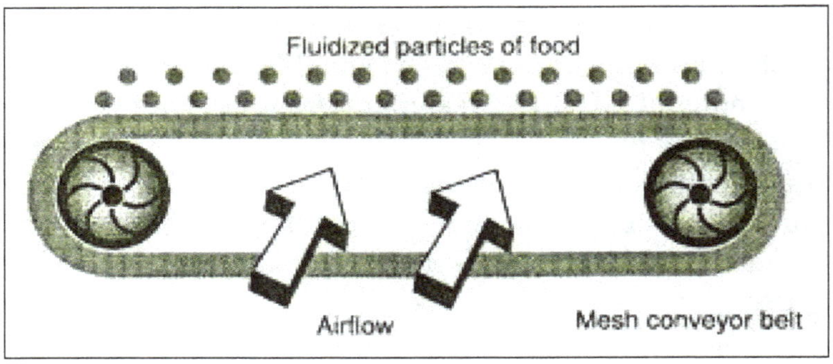

Fluidized-bed Freezer

In this type of air freezing, solid particles ranging in size from peas to strawberries are being exposed through a movement of the cold air (-18 °C to -34 °C) at high velocity as they pass along a conveyor belt. This will impart a vibratory motion to food particles, accelerating the freezing rate. The cold air being forced upward through the bed lifts and suspends the food particles, thus fluidization occurs. In this way, a rapid freezing rate is accomplished and an IQF (individually quick frozen) product is produced. In other words, food items are frozen as individual pieces and are not stuck together.

Advantages and Disadvantages of Air blast freezing; Fluidize-bed freezing (IQF):

- Advantages:
  - Economical.
  - Can freeze various sizes and shapes of food.
  - IQF has more efficient heat transfer, increased rate of freezing.
- Disadvantages:
  - Possible excess dehydration (freezer burn).
  - Undesirable bulging of the packages (by expansion of the product) may occur.
  - Non-uniform products cannot be fluidized (IQF) easily.

Indirect Contact Freezers

Indirect Contact freezers are used in the production of various frozen food commodities. In these freezers, food is placed on belts or trays and a refrigerant circulates through a wall beside the food. As the food comes into "contact" with the cold wall, it quickly cools down and freezes. Plate and slush freezers are some examples of indirect contact freezers.

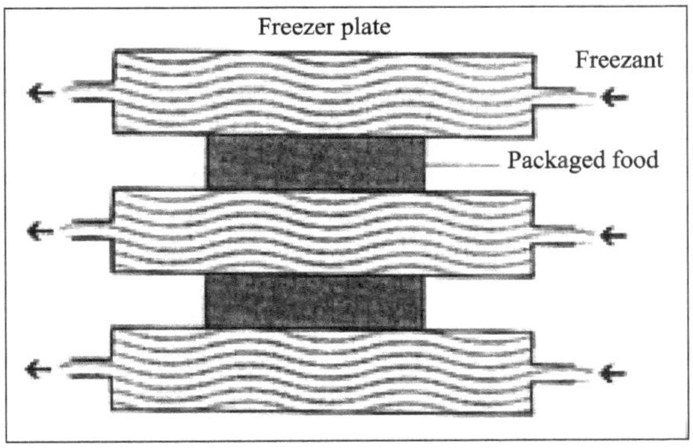

Plate Freezer.

During plate freezing food products are placed in contact with a metal surface which is cooled by a cold brine, or a vaporizer refrigerant such as ammonia. The packaged food either rests on, slides against or is pressed between the cold metal plates. These plates maintain firm contact with two major surfaces of packages to facilitate heat transfer and prevent bulging of the packages during the freezing process. Fish sticks and frozen fish fillets are commonly frozen in plate contact freezers.

Another type of indirect contact freezer is the slush freezers or scraped surface heat exchangers. These freezers can be used only for fluid food products. A common example of a scraped surface freezer is the machine used to convert ice cream mix to soft ice cream in restaurants and ice cream shops. The same principle is used in the commercial production of ice cream that is sold as hard ice cream. In the case of ice cream, the rotator not only aids in promoting rapid freezing and the development of small ice crystals, but it also aids in the incorporation of air bubbles into the freezing mix which results in the formation of a solid foam.

Advantages and Disadvantages of Indirect Contact Freezing:

- Advantages:
  - Economical.
  - Minimal dehydration.
  - Minimal package bulging.
- Disadvantages:
  - Slow freezing process.
  - Products must be of uniform thickness.

Immersion and Cryogenic Freezing

1. Immersion freezing involves the immersion of packaged or un-packaged food products directly in a non-toxic refrigerant fluid. The refrigerant fluids commonly used are propylene glycol, glycerol, sodium chloride, calcium chloride, and mixtures of salt and sugar. Canned citrus juice, turkeys and chickens are often frozen in immersion freezing units. Ice cream popsicles can also be frozen using this method.

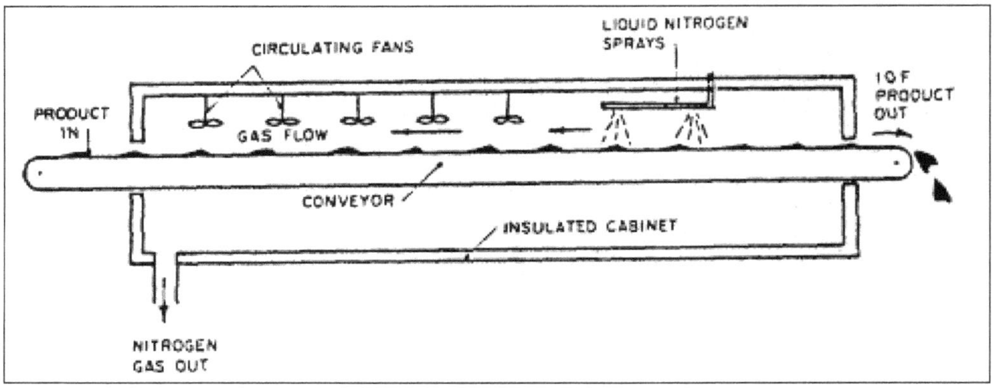

Cryogenic Freezing.

2. Cryogenic freezing is accomplished with cryogenic liquids, with liquid nitrogen being the most commonly used. This is a very rapid freezing method in which un-packaged or thinly packaged foods are exposed to extremely cold freezant. In contrast to the liquid immersion freezing, heat removal is accomplished during a change of state by the freezant. Products such as TV dinners, preformed patties and other high value food products are frozen in cryogenic freezers because of the excellent retention of quality imparted by the rapid rate of freezing and small ice crystal formation. The figure on the right illustrates a cryogenic process.

3. Liquid nitrogen cryogenic freezing: The product is first placed on a conveyor belt and is moved into the pre-cooling part of the freezing unit. Once the food is cooled, the food is sprayed by liquid nitrogen as it is being moved through the conveyor belt; here is where the freezing process takes place, by the nitrogen boiling as it contacts the food. Finally, the food is allowed to equilibrate to the desired final temperature (between -18 °C to -30 °C).

Advantages and Disadvantages of Immersion/Cryogenic Freezing:

- Advantages:
  - Rapid freezing process.
  - Almost no dehydration.
  - Oxygen is excluded, decreasing oxidative spoilage.
  - Individual freezing pieces have less freezing damage.
- Disadvantages:
  - Difficult to find suitable freezants.
  - Cost of operating is high.

Ohmic Heating and Inductive Heating

Ohmic Heating and Inductive Heating is also termed 'resistance heating' or 'electroheating', this is a more recent development in which an alternating electric current is passed through a food, and the electrical resistance of the food causes the power to be translated directly into heat. As the food is an electrical component of the heater, it is essential that its electrical properties (its resistance) are matched to the capacity of the heater. Heat treatment for complex food fluids is considerably improved when newer systems such as microwave heating, inductive or ohmic (or direct electrical) heating are used. These volumetric heating methods generate heat inside the food and depend less on thermal conduction and convection and so cause fewer temperature gradients.

Induction heating is the process of heating an electrically conducting object (usually a metal) by electromagnetic induction, where eddy currents (also called Foucault currents) are generated within the metal and resistance leads to Joule heating of the metal.

Ohmic heating has potential as an alternative heating technology because of its capability of delivering high quality food, rapid heating, compact design, and minimal heating of the product surroundings. Since heat is generated and distributed internally in the food, ohmic heating achieves close to 100% energy transfer efficiency and eliminates the need for an external heat transfer medium and surfaces that are associated with conventional heating technologies. Ohmic heating is more efficient than microwave heating because nearly all of the energy enters the food as heat. Another important difference is that microwave and radio frequency heating have a finite depth of penetration into a food whereas ohmic heating has no such limitation. However, microwave heating requires no contact with the food, whereas ohmic heating requires electrodes to be in good contact. In practice the food should be liquid or have sufficient liquid with particulate foods to allow good contact and to pump the product through the heater.

The advantages of ohmic heating are:

- The food is heated rapidly (1 $°Cs^{-1}$) at the same rate throughout and the absence of temperature gradients results in even heating of solids and liquids if their resistances are the same.
- Heat transfer coefficients do not limit the rate of heating.
- Temperatures sufficient for UHT processing can be achieved.
- There are no hot surfaces for heat transfer, as in conventional heating, and therefore no risk of surface fouling or burning of the product which results in reduced frequency of cleaning.
- Heat sensitive foods or food components are not damaged by localised over-heating.
- Liquids containing particles can be processed and are not subject to shearing forces that are found in, for example, scraped surface heat exchangers.
- It is suitable for viscous liquids because heating is uniform and does not have the problems associated with poor convection in these materials.
- Energy conversion efficiencies are very high (>90%).
- Lower capital cost than microwave heating.
- Suitable for continuous processing.

Foods that contain water and ionic salts are capable of conducting electricity but they also have a resistance which generates heat when an electric current is passed through them. The electrical resistance of a food is the most important factor in determining how quickly it will heat. Conductivity measurements are therefore made in product formulation, process control and quality assurance for all foods that are heated electrically. Electrical resistance of a food is measured using a multimeter connected to a conductivity cell. The most important feature of ohmic heating is the rate of heat generation, which in addition to the electrical resistance of the product, depends on the specific heat capacities of each component, the way that food flows through the equipment and its residence time in the heater. If the two components have similar resistances, the lower moisture (solid portion) heats faster than the carrier liquid.

Ohmic heating can therefore be used to heat sterilize particulate foods under UHT conditions without causing heat damage to the liquid carrier or over-cooking of the outside of particles. Furthermore, the lack of agitation in the heater maintains the integrity of particles and it is possible to process large particles (up to 2.5 cm) that would be damaged in conventional equipment. The resistance in an ohmic heater depends on the specific resistance of the product, and the geometry of the heater. The design of ohmic heaters must include the electrical properties of the specific product to be heated, because the product itself is an electrical component. Ohmic (or resistance) heating uses the electrical resistance of foods to directly convert electricity to heat. Ohmic heaters should therefore be tailored to a specific application and the following factors taken into account:

- The type of product (electrical resistance and change in resistance over the expected temperature rise.
- Flowrate.
- Temperature rise (determines the power requirement).
- Heating rate required.
- Holding time required.

To be commercially successful, ohmic heaters must:

- Have effective control of heating and flow rates and be cost effective.

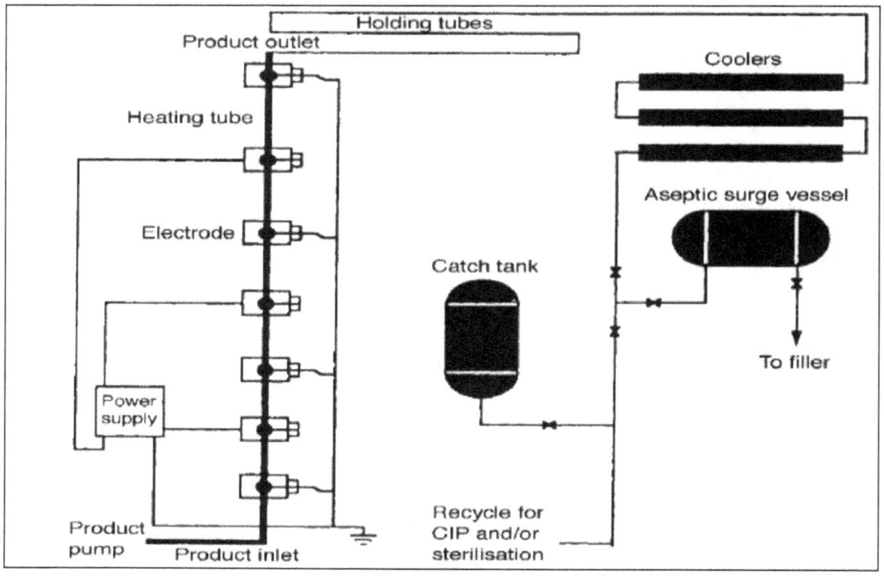

Ohmic heating.

Foods can be heated by either direct or indirect methods: dielectric and ohmic heating are direct methods in which heat is generated within the product, whereas infrared heating is an indirect method that relies on heat that is generated externally being applied to the surface of the food mostly by radiation, but also by convection and to a lesser extent, conduction. The main differences between dielectric, ohmic and infrared energy can be summarised as follows:

- Dielectric energy induces molecular friction in water molecules to produce heat, whereas ohmic heating is due to the electrical resistance of a food and infrared energy is simply absorbed and converted to heat.
- Dielectric heating is determined in part by the moisture content of the food, whereas the extent of heating by radiant energy depends on the surface characteristics and colour of the food and ohmic heating depends on the electrical resistance of the food.
- Dielectric and ohmic heating are used to preserve foods, whereas infrared radiation is mostly used to alter the eating qualities by changing the surface colour, flavour and aroma.
- Commercially, microwaves and radio frequency energy are produced at specified frequency bands that are allocated to prevent interference with radio transmissions, whereas radiant heat is less controlled and has a wider range of frequencies. Ohmic heating uses mains frequency electricity.
- The depth of penetration into a food is directly related to frequency; the lower frequency dielectric energy penetrates more deeply than radiant energy. In contrast, ohmic heating penetrates throughout the food instantly.
- The thermal conductivity of the food is a limiting factor in infrared heating, whereas it is not so important in dielectric and ohmic heating. Ohmic heating has been used to process various combinations of meats, vegetables, pasta and fruits when accompanied by a suitable carrier liquid. A variety of shapes, including cubes, discs, spheres, rods and twists have been processed. The almost complete absence of fouling in ohmic heaters means that after one product has been processed, the plant is flushed through with a base sauce and the next product is introduced. At the end of processing, the plant is flushed with a cleaning solution. In conventional heaters, turbulence is needed to create mixing of the product and maintain maximum temperature gradients and heat transfer coefficients. In ohmic heating, the electric current flows through the product at the speed of light and there are no temperature gradients since the temperature is uniform across the cross-section of flow. For the successful applications of ohmic heating, careful formulations of the food mixture, in terms of electrical and rheological parameters, should be carried out in order to prevent particle/liquid slip for the same electrical conductivity. Inductive heating is restricted to heating metal plates, tubes, screws and coils and the industrial applications are very limited (extrusion cooking, cookers for caramel, hobs) and information on them extremely limited. The volumetric heating is low, with a poor energy conversion. Ohmic heating technology is considered a major advance in the continuous processing of particulate food products. Ohmic heating is direct resistance heating by the flow of an electrical current through foods, so that heating is by internal heat generation. One advantage of ohmic heating is that large sized heating tubes with lower shear rates can be used. This allows heating of fragile particles.

Basic Principles of Ohmic Heating

The basic principle of ohmic heating, or Joule effect, is the dissipation of electrical energy in the form of heat using an electrical conductor. This energy generation is proportional to the square power of the local electric field strength and the electrical conductivity. The power generated, ΔP, in a homogeneous element conductor of given length z in the z direction and area ΔA, perpendicular to the electrical field E, with constant physical properties.

Product Suitability–Formulation and Pre-Treatment

Ohmic heating is particularly suitable for viscous products and foods containing particulates because the heat is generated inside the food product. The critical property influencing the rate of ohmic heating is the electrical conductivity, which depends on a number of factors including temperature, ionic constituents, material microstructure and field strength. As most solid vegetable particles have lower electrical conductivities than liquids, particular care must be taken in the design of formulations to eliminate temperature changes. One way to obtain solid/liquid heating homogeneity is to use a pilot-scale batch ohmic heater and to seek the optimum salt concentration within the liquid phase, thereby minimizing the average difference in heating rate between solid and liquid. The closer the match between the electrical conductivities of the different phases, the more even the current flow will be; this still may not lead to uniform heating because the thermal capacities of the phases may be different and the temperature dependence of the thermal conductivities may also differ.

Thus food treatment by ohmic heating often needs a pre-treatment in aqueous NaCl solution to increase the electrical conductivity of the solid phase. Also, high temperature short time processing of vegetables requires a blanching pre-treatment to degas and inactivate enzymes. It is well known that fruit and vegetable texture is profoundly altered by thermal processing; firmness decreases as temperature increases. However, a low temperature long-duration pre-treatment can improve the final texture of treated vegetables. The mechanism claimed to explain the firming effect is the demethoxylation of pectic materials of the cell wall catalysed by the pectin methylesterase. Thus a combination of low temperature longduration pre-treatment and salted water offers an opportunity to improve the electrical conductivity and the final quality of the product after ohmic heating processing, even for fragile products. Eliot et al., Eliot and Goullieux have evaluated and optimized low temperature long-duration pre-treatment before ohmic heating in a batch unit with potato, a vegetable commonly used in prepared meals, and cauliflower, a brittle product that does not resist conventional thermal treatments.

- In the case of potato cubes, a significant loss compared with fresh product after conventional blanching at 95 °C, but a better firmness after pre-treatment at 60 °C and 55 °C (the improvement reached up to 90 per cent for 55 °C versus fresh).
- In the case of cauliflower florets, a significant loss was observed compared with fresh product after conventional blanching and a low fall in firmness after pre-treatment at 50 °C and 40 °C.

After pre-treatment and cooling, potato cubes and cauliflower florets were placed in an ohmic heating pilot plant successively reproducing the heating, holding and cooling stages of an ohmic heating process. Heating was obtained by a steady electric field (between 1 and 30 V/cm) from a 50 Hz AC 5 kW power set connected to a rototransformer. The volume of the batch was made up to 1l with a NaCl solution (0.5 g/kg for the cubes, 5 g/kg for the florets) to maintain the homogeneity of the electrical conductivity of the vegetable mixture/NaCl solution (0.40–0.45 S/m) and to apply ohmic treatments with similar heating rates. The mixture was then heated to 135 °C and held at this temperature for 30s. After ohmic heating, the final quality was again evaluated by mechanical firmness measurements. Potato cubes pre-treated at 60 °C exhibited higher firmness, although there was no significant difference from pre-treatment at 55 °C. The textural quality of cubes pre-treated at low temperature was better than those of cubes pre-treated at 95 °C. An increase of 77.6 per cent and 100 per cent was observed for the cutting work, respectively at 60 °C and 55 °C. The cutting work decreased between the first thermal treatment and the ohmic heating. The most important drop was measured for samples pre-treated at 55 °C and 60 °C, with a value of the parameter reduced by a factor of 4–5, when they had undergone an increase in firmness during pre-treatment.

The value of this parameter was only halved for blanching at 95 °C. However, compared with fresh product, cubes pre-treated at low temperature and ohmically heated held 35–40 per cent of their initial firmness versus 20 per cent for conventional blanching. The ability of ohmic heating to rapidly generate heat inside products can be used to accelerate cooking. The fast food sector usually cooks patties frozen to prevent bacterial contamination during thawing. The long time needed to reach 70 °C at the centre is due to the high latent heat of ice melting. Consequently, this cooking method entails high power consumption, alters organoleptic properties and does not guarantee the required kill level for all pathogenic microorganisms. Özkan et al. reported a new method of hamburger patty cooking combining ohmic and plate heating. At 50 V AC, the required cooking was reduced by around 30 per cent, while the taste and texture were not affected. The authors concluded that this combined ohmic and plate heating was very successful in reducing cooking time and producing safer products. Thawing is a critical operation for the microbial quality of a foodstuff. It must be carried out rapidly at low temperature. The conventional process is to place the frozen product in circulating water (temperature below 20 °C), thereby reducing moisture loss. However, there are several disadvantages: possible microbial growth on the product surface, reduction of nutrient quality due to leaching of soluble proteins, high consumption of fresh water and consequently large quantities of loaded wastewater. Dielectric or microwave thawing avoids these problems, but the product surface may be cooked before its centre is thawed. Liquid water absorbs microwaves more rapidly than ice and so the peripheral front of thawing prevents microwave penetration in the non-thawed central region.

An intermittent exposure, allowing temperature stabilization by thermal conduction, is usually applied. Ohmic heating is another alternative to thaw frozen food blocks, but its adoption is hindered by hot spots. Electrical conductivity increases with temperature and is approximately two orders of magnitude lower for frozen food than for thawed food. Thus the thawed portion of a block can cook while the rest is still frozen. Roberts et al. developed an ohmic thawing unit with surface temperature sensing and computer-automated capabilities that can thaw shrimp blocks without runaway heating problems. The ability of liquid-contact thawing of frozen meat by ohmic heating was reported by researchers. The latter used brine as a carrying fluid and found that the thawing process was faster when the brine concentration increased and the largest surface of the sample was perpendicular to the electrical field. No significant changes in sample colour, pH or brine electrical conductivity were observed after ohmic thawing. Therefore ohmic thawing is a rapid method worth developing in the future. Ohmic heating is an alternative method that has been studied for vacuum evaporation of orange juice by Wang and Chu, in which they showed that vacuum evaporation by ohmic heating could evaporate more moisture than a conventional process in the same time. Moreover, the final product was brighter and kept more aroma. Application of ohmic heating for pasteurization, HTST sterilization, cooking, thawing, blanching and evaporation of food products mainly uses its thermal contribution. However, the non-thermal contribution of ohmic heating also has wide applications.

The influence of ohmic heating on mass transfer can also be useful to enhance water removal. The hot-air drying rate can be significantly increased during most of the drying process by low frequency and high electric field strength. In addition, Wang and Sastry have observed a shift of sorption isotherms for ohmic pre-treated samples, indicating a permeabilization of the structure and a redistribution of water within the vegetables. The drying processes are time and energy intensive and so ohmic heating has been studied as a possible accelerating pre-treatment. Zhong and Lima obtained a 24 per cent decrease in vacuum drying time with minimal ohmic heating (50 V/cm, 45 °C) but at final moisture level no difference from that of raw sweet potatoes. In this case, the mild electroporation known to occur during ohmic heating was not evident. The modelling of ohmic treatment of solid/liquid mixture shows the effect of slip velocity, solid fraction and particle and tube size. The flow behaviour of food mixtures becomes a critical factor as (i) the residence time distribution of particles and liquid, (ii) the consequences for the outlet temperature of liquid and solid and (iii) the product quality. The prediction of particle passage times is required to ensure sterility, cooking and optimized product quality. It is important to be able to predict minimum and maximum passage times of particles.

Technical Considerations

Though technologically simple, ohmic heating has some technical limitations in its electrical, mechanical and hygiene design. As the electrodes are in direct contact with the product, consequently from the food and hygiene viewpoint, corrosion has to be prevented or remain quantitatively and compositionally harmless with respect to the product. Unlike pulsed electric fields, in ohmic heating devices the signal is always an alternating, bipolar (and symmetrical) current. At a low frequency, partial electrolysis of the solution and electrode corrosion have to be prevented in food treatment. These faradaic reactions are generated through electrode/product interfaces and enhanced by a low frequency periodic signal, high current density, inappropriate electrode material, high temperature or aggressive product. To prevent electrolysis at low frequency (50 or 60 Hz) alternating current, Stirling and Berthou et al. use titanium electrodes coated with platinum or ruthenium and limit the current density to below 5000 A/m2. Consequently, the electrodes used are expensive; but the power supply is a conventional low-cost transformer. The high frequency limits electrolysis especially for high current densities on the electrodes. Emmepiemme – Italia and Reznik use frequency up to 100 kHz with conventional stainless steel or graphite electrodes. The signal is sinusoidal or rectangular in shape and the power is supplied on the signal fraction. This power supply is an electronic transformer. Reznik adds an electrolyte between the electrode and the product to prevent product contamination.

Radio Frequency Processing of Food

Radio frequency (RF) heating is a commonly used food processing technology that has been applied for drying and baking as well as thawing of frozen foods. Its use in pasteurization, as well as for sterilization and disinfection of foods, is more limited.

The Basic Science

RF processing uses dielectric heating to thermally process foods using electromagnetic waves. RF wavelengths cover the range of the electromagnetic spectrum from 3 kHz to 300 MHz. Only frequencies of 13.56 MHz, 27.12 MHz, and 40.68 MHz are permitted for use in the United States. During RF processing, dielectric materials are placed in an alternating electric field. RF energy is generated by a triode valve and applied to the food via a pair of electrodes. Dielectric heat energy is generated in the food by molecular friction as high-frequency electric fields alternate. Unlike conventional heating, dielectric heating is fast and volumetric.

RF processing can be used to process any dielectric material and most food products are dielectric. Foods contain polar molecules such as water, and these molecules have dipole moments, positive and negative charge centers that do not coincide. When placed in an electric field, polar molecules align with the field via the dipole moment phenomenon. In this way, polar molecules rotate continuously to align with the field as the field is alternated. Friction is created between the molecules, converting electromagnetic energy into heat and increasing the temperature of the food. Dissociative ions in foods also produce heat through ionic conduction whereby the ions oscillate forward and backward in the food creating friction and heat.

The electromagnetic characteristics of the food are important in determining process parameters. The two important properties are permeability and permittivity. Permeability has a very small contribution to dielectric heating and is therefore not usually considered. Permittivity is the parameter most commonly used to describe the dielectric properties of foods. It relates the reflection of electromagnetic waves at interfaces and the attenuation of the wave energy within the food. The three most popular methods to measure dielectric properties are transmission line, open-ended coaxial probe, and the resonant cavity method. It is important to remember that dielectric properties are dependent on temperature and frequency as well as the density, structure, composition, and moisture content of the food.

Radio Frequency vs. Microwave

RF heating has several advantages over microwave heating. RF heating is simpler than microwave heating in that it generates an electrical field between a pair of electrodes in one direction in contrast with microwave heating, which transmits in all directions. Microwave equipment also requires a waveguide to transmit microwaves. In these ways, construction of large RF heating systems is simpler and its application to continuous processes more straightforward. Another difference is that RF processing uses lower frequencies than microwave, resulting in longer penetration depths that make it useful to process bulk food materials that have larger dimensions. RF processing also has the ability to dynamically control final product moisture content. Oscillator circuits can be dynamically matched to load automatically, controlling the power absorbed by the load. In this way, RF driers can automatically draw more power if the moisture in the product is too high. Finally, RF heating provides more uniform field distributions than microwave heating and as a result is slightly more efficient.

Equipment Design

There are two design types of RF heating equipment. They are the free-running traditional oscillator system and the Crystal Oscillator Source Matched Impedance Generator (COSMIG) 50 ohm system. Traditional oscillator-based RF systems are simple to use and lower in cost than newer COSMIG 50 ohm systems. The newer systems, although more costly, allow for precise frequency and power control with feedback on forward and reflected power, thus enabling precise control of processing parameters such as power and heating rate to obtain exact final product properties such as moisture content.

Parallel plate "through-field" RF systems are the most commonly used in the industry; however, there are two other configuration types. "Fringe-field" and "staggered through-field" systems, which contain rod- or tube-shaped electrodes, can be used to treat thin or intermediate thickness products.

Commercial Applications

Drying and Baking

Post-baking RF drying is common in cookie, cracker, cereal, and snack food manufacturing. Surface cracking often occurs during traditional drying processes due to moisture gradients in the product. RF drying solves this issue because the energy penetrates evenly, minimizing moisture gradients. The use of RF also avoids discoloration and flavor damage due to thermal buildup. It enables faster line speeds, more consistent moisture levels, lower drying temperatures, and smaller equipment footprints. In addition, RF is highly energy efficient for removing final moisture from dry baked products.

Traditional baking is carried out in forced-air convection ovens where the heat is transferred to the surface of the dough by convection, conduction, and irradiation, and then from the surface toward the inside by heat conduction. The dough has low heat conductivity; therefore, the traditional baking process is slow and can require as long as 30 min. In RF processing of bread, dough pieces are placed inside pans made of heat-resistant, RF compatible, food-grade polymers and introduced into the RF baking oven with controlled-temperature air flow and exhaust fans to evacuate excess moisture released during the baking process. Total baking times are reduced to 8–10 min for commercial bread loaves, and the bread does develop a crust when it comes out of the RF oven. Other techniques, such as infrared, can be used to form crusts if desired. In breads, RF processing has also been shown to reduce mold growth and staling during storage.

Defrosting

Traditional defrosting methods suffer from intrinsically slow heat transfer. The larger the size of product, the longer the time needed. This provides opportunities for bacteria to grow, results in high drip losses, and causes adverse changes in the product quality, leading to significant economic loss.

RF rapidly generates heat volumetrically, and defrosting is achieved in minutes rather than hours/ days, even for large product blocks. It can also be performed directly inside packaging. The heating process results in a significant reduction of drip losses while minimizing product deterioration due to bacterial growth, making it ideal for many defrosting processes.

Sterilization and Pasteurization

Sterilization processes like retorting are generally carried out by treating products with drastic heat treatments, resulting in loss of product quality. Pasteurization is a relatively gentle heat treatment carried out at temperatures below the boiling point of water. Both traditional sterilization and pasteurization methods use saturated steam or pressurized hot water to heat the packaged food. The temperature rise of the product is slow and non-uniform, frequently resulting in overheating of the outside layers of the food, which results in quality losses. The ability of RF to rapidly generate heat volumetrically within the product minimizes product deterioration. The RF equipment requires less space compared to other machines based on conventional heating technology that have long heating times and require large product volumes under processing at any given time. Because no thermal energy is wasted into the environment, the RF process is more efficient, economic, and environmentally friendly. RF pasteurization is used in meat processing. Liquid pasteurization applications are also being explored.

One of the most prevalent applications for RF pasteurization is in the treatment of dry ingredients. In a dry state (5% and 15% moisture content) microbes are considered "dormant" and are difficult to kill. RF is very effective with dry ingredients like flour, cereal grains, protein supplements, spices, seeds, and pet foods, insuring their food safety. This is particularly important for ingredients that are added to foods after the cooking "kill step," such as cookie dough inclusions for ice creams and cold-pressed health bars.

Disinfestation

Insect infestation is another major problem. Chemical fumigants (like methyl bromide) are commonly used for postharvest pest control, but they have environmental concerns and regulatory issues. Alternative thermal treatments frequently result in product quality deterioration. RF processing allows for rapid and uniform heating of many substrates. Recent research has confirmed the viability of RF for disinfestation of fruits and nuts.

RF and the Future

Continued expansion of applications of RF processing of foods is anticipated during the years ahead. Ongoing research on dielectric properties of foods (including heterogeneous foods), final RF-treated food quality and safety, and advances in mathematical modeling all support increased implementation of RF technology and additional FDA approvals of RF processing for specific food applications like sterilization. Challenges related to avoiding dielectric breakdown (arching) and thermal runaway heating from hot spots are being addressed by researchers and equipment manufacturers. And specific new packaging materials tailored for RF processing continue to be developed.

Microwave Applications in Food Processing

Microwaves are defined as a part of electromagnetic waves which have frequency range between 300 MHz and 300 GHz corresponding to wavelength from 1mm to 1m. Microwaves heating can be applied in a broad range of food processing such as drying, tempering, blanching, cooking, pasteurization, sterilization, and baking. Microwave heating has considerable advantages over conventional heating.

In a conventional oven, a gas or charcoal fire, or an electric heating element, generates heat outside of the material by convection and inside by of the material by conduction. In microwave heating, by contrast, the inside of the material is heated first. The process through which this occurs primarily involves the conduction losses in materials with large loss tangent.

An interesting fact is that the loss tangent of many foods decreases with increasing temperature, so that microwave heating is to some extend self-regulating. This result is that microwave cooking generally gives faster and more uniform heating of food, as compared with conventional cooking. The efficiency of a microwave oven, when defined as the ratio of power converted to heat (in the food) to the power supplied to power supplied to oven, is generally greater than the cooking efficiency of a conventional oven.

A common misconception is that microwave heating is always more expensive than heating by conventional techniques. This will actually depend on the application and utility costs. However, in some cases, microwaves can be 50% or more efficient than conventional systems, resulting in major savings in energy consumption and cost.

Methodology

Qualitative sourcing of information from the library and online surfing was adopted in the preparation of this paper based on the emerging development in microwave heating techniques in food processing.

The Principles of Microwave Heating

Dipolar Interaction

Once microwave energy is absorbed, polar molecules such as water molecules inside the food will rotate according to the alternating electromagnetic field. The water molecule is a "dipole" with one positively charged end and one negatively charged end. Similar to the action of magnet, these "dipoles" will orient themselves when they are subject to electromagnetic field. The rotation of water molecules would generate heat for cooking. Hence the composition of a food will affect how it will be heated up inside the microwave oven. Food with higher moisture content will be heated up faster because of the dipolar interaction.

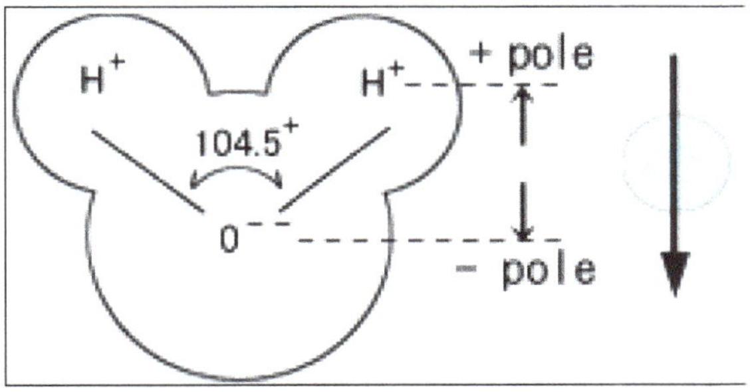

The structure of water molecules (left) and Image of permanent dipole (right).

Then as shown in figure, when there is no external electric field, it has set a balance. But when placed in external electric field, dipole will turn towards the electric field.

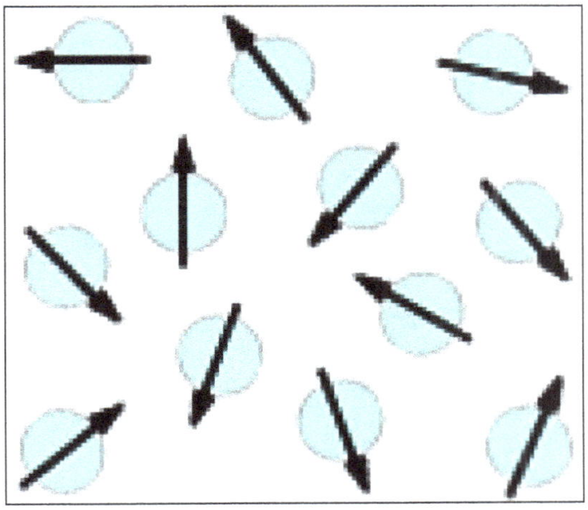

When there is no external electric field.

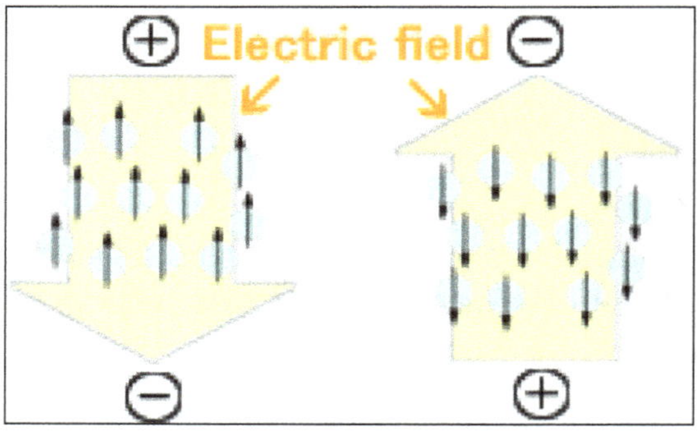

When there is an external electric field.

Figure show the Direction of dipoles influenced by external electric field. Figure above shows a case where a too much lower frequency of radio wave is irradiated to the permanent dipole of water. In this case, the permanent dipole will immediately follow the directions of electric field. So in this case, water doesn't generate heat. On the other hand, Figure shows a case where a too much higher frequency of radio wave is irradiated to the permanet dipole. In this case, since electric field changes its direction too fast, dipole won't be able to follow. Then, water does not generate heat in this case also.

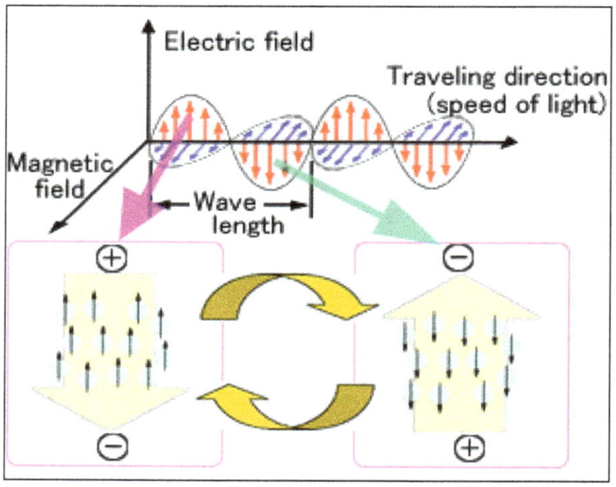

At too lower frequency of radio wave.

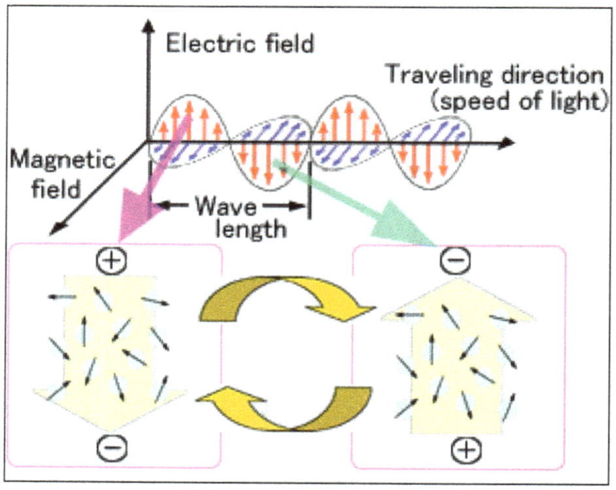

At too higher frequency of radio wave.

In contrast to these, Figure shows a case where moderate frequency of radio wave is irradiated to the permanent dipole. In this case, the permanent dipole changes a little behind the electric field. During the time delay, water is absorbing energy from radio wave and generates heat. And, this moderate frequency is the microwave.

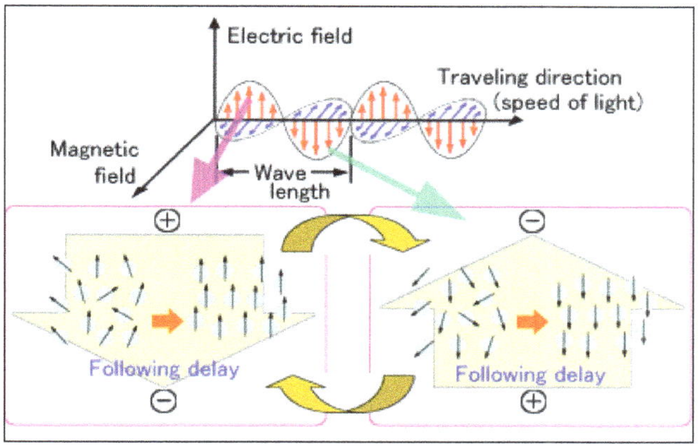

At moderate frequency of radio wave (microwave).

The Basic Structure of Microwave Heating Device

Below, a microwave oven is relatively simple system consisting of a high power source, a wave guide feed, and the oven cavity. The source is generally a magnetron tube operating at 2.45GHZ. Although 915MHZ is sometimes used when greater penetration is desired. Power output is usually between 500 and 1500W. The oven cavity has metallic walls, and is electrically large. To reduce the effect of uneven heating caused by standing waves in the oven, a mode stirrer, which is just a metallic fan blade, is used to perturb the field distribution inside the oven with a motorized platter. The food is also rotated with a motorized plate.

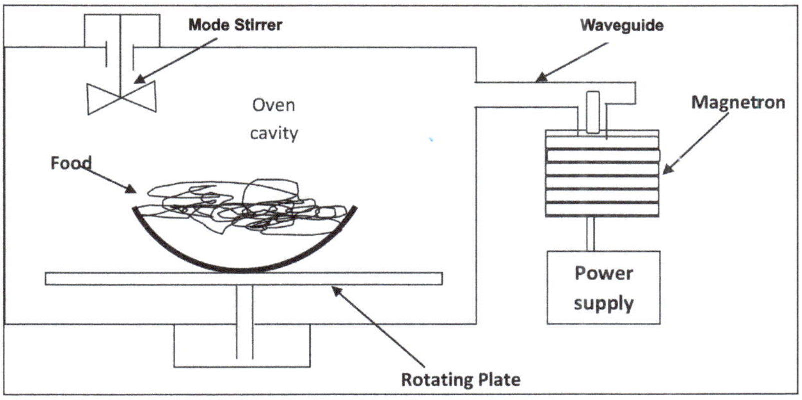

Structure of Microwave Heating Device.

Functional Parts of the Microwave Heating Device

1. Generator (Magnetron): Magnetron is a device built inside the Generator that oscillates microwave. The microwave that is oscillated by the Generator is called the traveling wave (or incident wave).The theory of operation of the magnetron is based on the motion of electron under the influence of combined electric and magnetic fields. The direction of the electric field is from the

positive electrode to the negative electrode. The magnetron can produce oscillations at microwave frequency by virtue of the current induced electrostatically by moving beams of electrons. This frequency is determined by the type of cavities used. The microwave which is generated by magnetron propagates in waveguide to be emitted from the nozzle.

2. Isolator: This is device, sends traveling wave directly to the Applicator, and absorbs reflected wave by built-in dummy load to avoid returning back to the Generator. Reflected wave which occurs by rotation of the Turntable and the Stirrer. Isolator can reduce the influence of reflected wave fluctuation. Without this fluctuation, magnetron can continue stable operation. In other words, Isolator functions to protect magnetron.

3. Power Monitor: This is a device that monitors traveling and reflected waves of microwave power propagating through the rectangular waveguide. Must be careful when reflected wave becomes large, there is an increase in error.

4. Waveguide: Waveguide is a hollow metal pipe use to conduct electromagnetic waves through its interior. A wave guide can be of any shape. The most popular shape is rectangular, circular and even more exotic shapes. Microwave (electromagnetic wave) propagates in the interaction of electric and magnetic fields. Microwave is transmittable when metal pipe with cross section is used. In general, for the microwave heating equipment, 2GHz standard rectangular waveguide of rectangular cross section is used. (Standard waveguide: WRJ-2/WRI-22, Flange: BRJ-2/FUDR22).

5. EH-Tuner: There are two kinds of tuners, which are three stub and EH. By adjusting E- or H-tuner, changes the phase and magnitude of microwave reflection at the tuner section. It is also adjustable to set the display value of reflected power to zero by adjusting E- or H- tuner. This means that, by adjusting E- or H- tuner generates a same magnitude reverse phase wave to counter the reflected wave. And as a result, the reflected wave has been denied. When the reflected power wave value is zero on the display of Power Monitor, power consumption of after tuner to inside applicator is maximized. This condition is called "The Matching".

6. Applicator: A heating tank that heats the object placed inside by irradiation of microwave. Depending on the application, there are a variety of shapes, such as batch type, conveyor type, waveguide type, etc.

However, the most critical issue in the design of a microwave oven is safety. Since a very high power is used, leakage levels must be very small to avoid exposing the user to avoid exposing the exposing the user to harmful radiation. Thus, the magnetron, feed waveguide, and oven cavity must all be carefully shielded. The door of the oven requires particular attention; besides close mechanical tolerances, the joint around the door usually employ Radio frequency (RF) absorbing material and a $\lambda/4$ choke flange to reduce leakage to an acceptable level.

The Characteristics of Microwave Heating

There are some characteristics of microwave heating that other heating methods (conventional heating method) do not have. The following are the features:

- The Internal Heating: Microwave energy will reach the object to be heated at the same speed of light. Then it enters into the object as a wave, and by getting absorbed, the object generates heat. Therefore as shown in Figure, microwave heating is internal heating.

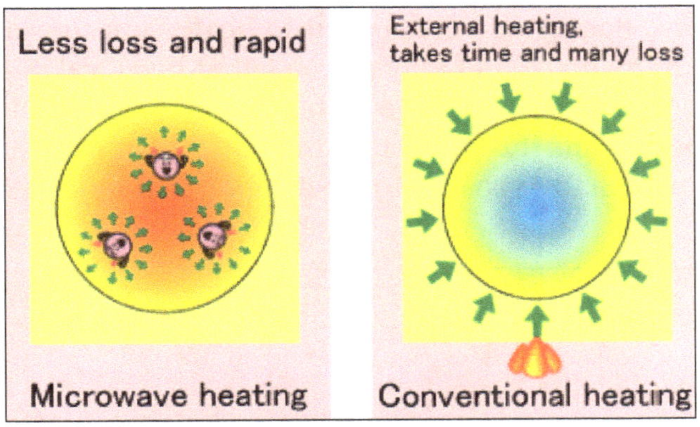

Microwave heats object internally.

- Rapid Heating: As shown in Figure, in conventional heating, the object's temperature rises by spreading heat energy from the surface to inside (external heating) On the other hand, by microwave heating, the object will generate heat on their own by the penetration of the microwave. Not necessary to consider about the heat conduction. That is why rapid heating is possible by microwave. Although the object has to be large enough for the microwave to penetrate, the smaller objects will also be heated from inside as the depth of microwave penetration.
- High Heating Efficiency: Microwave penetrates into the object at the speed of light. You get high heating efficiency because no need to consider the heating losses of air inside the heating furnace.
- Rapid Response and Temperature Control: Microwave penetrates into the object at the speed of light. So it allows rapid response. For example, you can start and stop the heating instantly. In addition, by the adjustment of microwave output, you can control the amount of heat energy generated inside the heated object.
- Heating Uniformity: Each part of the heated object generates heat, so even for those objects with complicated shape, it can be heated relatively uniform. To keep the heating uniformity, stirrer, turntable, and belt conveyor is used for heating blur related to wave length.
- Clean Energy: Microwave doesn't require a medium, because it propagates only by changes of electric fields and magnetic fields. It can propagate in a vacuum. It reaches the object and penetrates without heating the air. The heated object generates heat by absorbing microwave energy to convert it to heat energy. Therefore, it can be said as clean energy because it doesn't heat the air during process.
- Good Operation and Work Environment: Conventional heating requires a heat source, and the temperature rises not only of heated object, but also of heat source and the heating furnace. So the temperature of room equipped heating furnace goes high because of radiant heat. This is an operation and work environment issue. On the other hand, microwave heating only uses electricity to generate heat of the object. The temperature of the object only raises, not the furnace. And there is no radiant heat, so it's possible to keep operational and good working environment.

Biological Effects and Safety

The proven dangers of exposure to microwave radiation are due to it thermal effects. The body absorbs RF and microwave energy and converts it to heat, as in the case of a microwave oven this heating occurs within the body, and may not be felt at low level. Such heating is most dangerous to brain, the eye, the genitals and the stomach organs. Excessive radiation can lead to cataracts, sterility or cancer. Thus it is important to determine a safe radiation levels. The most recent safety standard for human exposure to electromagnetic field is given by IEEE Standard C95.1-1991.

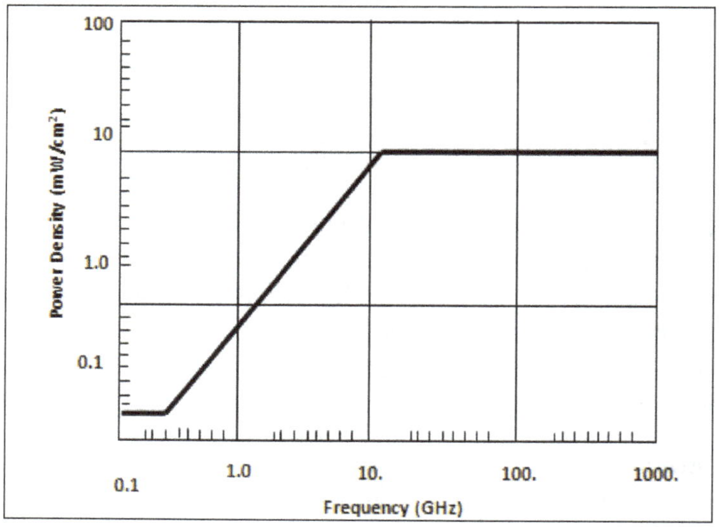

IEEE Standard C95.1-1991recomended Power density limits for human exposure to RF and microwave electromagnetic fields.

In the RF microwave frequency range of 100MHZ to 300GHZ, exposure on the limits are set on the power density (W/cm^2) as frequency, as shown in figure. The recommended safe power density limit is as low as $0.2mW/cm^2$ at the lower end of this frequency range, because fields penetrate the body more deeply at lower frequencies. At frequencies above 15GHZ the power density limit rises to $10mW/cm^2$, since most of the power absorption at such frequencies occurs near the skin surface. By comparison, the sun radiates a power density of about $100mW/cm^2$ on a clear day, but the effect of this radiation is much less severe than a corresponding level of lower frequency microwave radiation because the sun heats the outside of the body, with much of the generated heat reabsorbed by the air, while microwave power heats from inside the body. At frequencies below 100MHZ electric and magnetic fields interact with the body differently than higher frequency electromagnetic fields and so separate limits are given for field components at these lower frequencies.

Food Extrusion

Extrusion is defined as a process in which material is pushed through an orifice or a die of given shape. The pushing force is applied using a piston or a screw. In food applications, screw extrusion is predominant. Historically, one can trace the use of a screw as a conveying device to the Greek philosopher Archimedes, who used a single screw in a cylindrical open channel to pump water uphill. Today's extruder consists of one or more screws encased in a metal barrel, attached to a drive motor. A hopper at one end is used to feed raw materials, while a die on the other gives shape to the product. Extruders were developed in the 1870s to manufacture sausage. Application of the single-screw extruder evolved during the 1930s, when it was used to mix semolina flour and water to make pasta products. It was also used in the process of making ready-to-eat (RTE) cereals to shape hot, precooked dough. In both of these applications, the level of shear rate was low. During the late 1930s and 1940s, directly expanded corn curls were made using extruders, which were characterized by extremely high shear rates. The first patent on an application of twin-screw extrusion technology was filed in the mid-1950s. Since then, the application of extrusion technology has widened and grown dramatically.

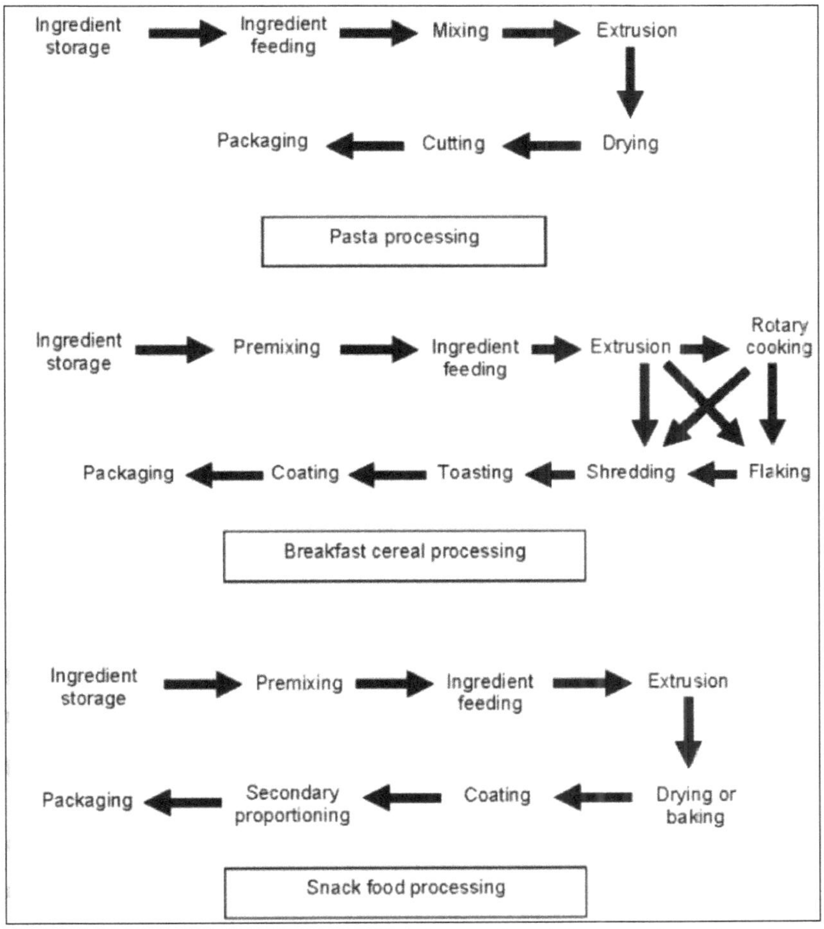

Role of extrusion in manufacturing of pasta, breakfast cereal, and snacks.

Extrusion processing of food materials has become an increasingly important manufacturing method, and its applications have broadened substantially in the last two decades. From the point of view of transport phenomena, extrusion processing can be thought of as a combination of several processes, including fluid flow, heat and mass transfer, mixing, shearing, particle size reduction, melting, texturizing, caramelizing, plasticizing, shaping, and forming. Depending upon the product, one or many of these processes will take place in an extruder. For example, in pasta manufacturing, the main objective of the extrusion process is to partially gelatinize starch, compact the dough, and give it the desired shape. In the case of chocolate manufacturing, the extruder is used as a reactor to generate key flavor attributes. In the case of corn puffs, the extruder is used to develop the desired expanded, porous structure. Today, a variety of products such as breakfast cereals, pasta, snacks, candy, confectionery products, and pet foods are made using screw extrusion processes. Figure shows the key role of extrusion in the manufacture of pasta, breakfast cereals, and snack products. An extrusion facility consists of many components, as shown in figure.

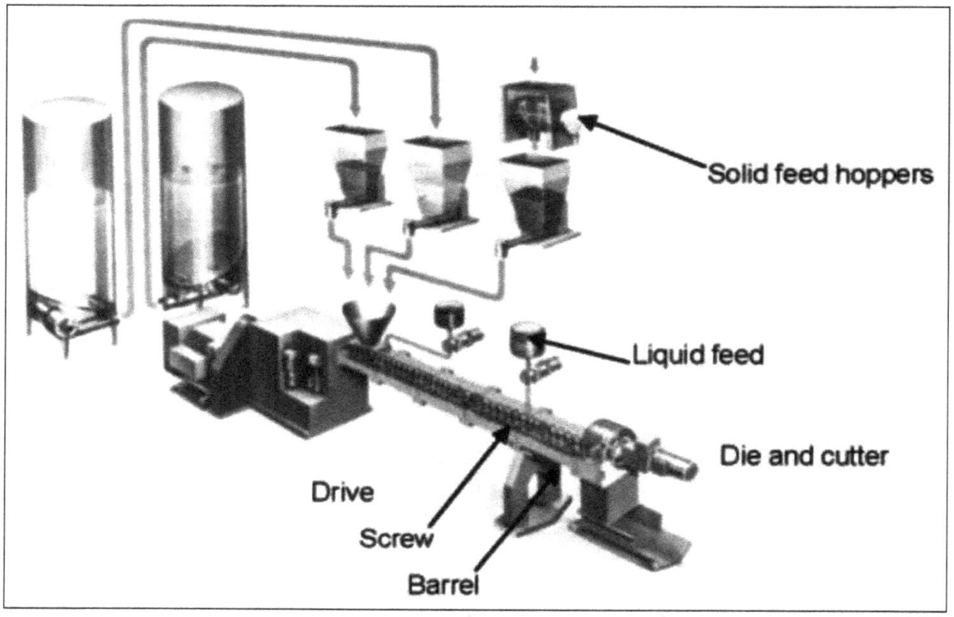

Typical twin-screw extrusion facility.

There are many advantages of extrusion processing over other conventional cooking processes. Extrusion is a continuous process. It is flexible because on-line process adjustments can be made to achieve desired product characteristics. In addition, the same extruder can be used to manufacture different types of products. The process has no effluents and is energy-efficient. It can be used to process relatively dry, highly viscous materials as well as moist or wet materials. Recent advances in the basic research and development of better process control strategies have made it possible, to a large extent, to control the thermo-mechanical changes during extrusion to achieve desired product properties.

Food extrusion is a high-temperature–short-time (HTST) process. Figure shows a comparison of the food extrusion process with other processes, on a diagram with temperature and residence time as axes. It can be seen that during extrusion, the food material may be exposed to temperatures as high as 200 °C. However, the exposure to such high temperatures lasts only 1–10 seconds. In a typical corn cooker, the operating temperature is about 100 °C, and exposure time is about an hour. Figure shows a flow diagram for the production of cornflakes using the conventional process and using an extruder. One can see that several steps that are used in conventional cooking are eliminated by using an extruder. Though there are some differences in the products obtained from the two processes shown in Figure, several manufacturers have adopted the extrusion process because of the advantages.

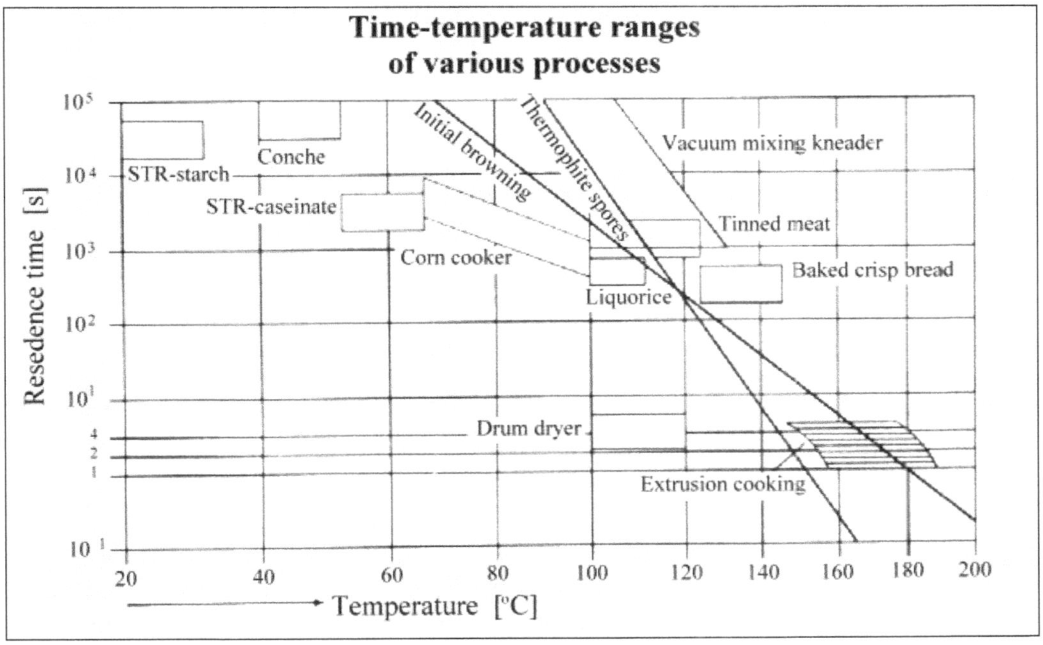

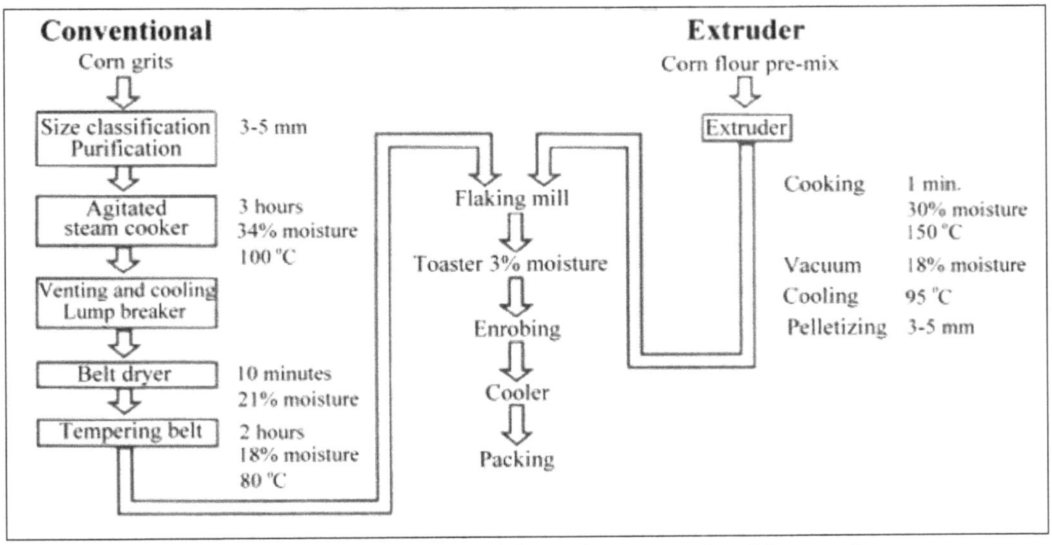

Flow diagram for the production of corn flakes using conventional & extrusion processes.

Extruded Products

Today, extruders are used in a variety of applications, such as the manufacture of breakfast cereals, pasta, meat analogs, filled snack products, and pet food. They are also used in the mechanical de-boning of meats, as well as hops processing. Extruded products can be broadly categorized into products for human consumption, products for animal consumption, and biodegradable, non-consumable materials.

Products for Human Consumption

Some examples of extruded products for human consumption are listed in table.

Table: Extruded products for human consumption.

Type	Example
Directly expanded	Breakfast cereals, corn curls
Unexpanded	Pasta
Half-products	Potato pellets
Co-extruded	Fruit-based cereal, jelly-filled cores
Modified	Starches, fat mimics
Texturized	Meat analogs
Candy	Licorice, chewing gum

For directly expanded products, extrusion is carried out under high shear, high temperature, and low moisture conditions. For pasta, extrusion is carried out at moderate (~40 percent w/w) moisture and a low (~50 °C) temperature. In half-products such as potato pellets, an extruder is used to make a dense, semi-cooked product, which is stored under appropriate conditions in a warehouse for further processing as needed. Because of its high density, a half-product occupies less space than a puffed product. These pellets are processed further, when required, to make puffed or expanded product by frying or baking. To make co-extruded products, a special die is designed to pump semi-solid material into the core. Modified starches and fat mimics are made by extruding under optimum time, temperature, and shear conditions. To make texturized meat analogs, plant proteins such as soy proteins are used along with a long die to impart a fibrous, meat-like structure to the extrudate.

Pet Food and Animal Feed Products

Extrusion cooking was used to manufacture pet foods during the early 1950s. Today, extrusion is used to make semi-moist and dry expanded pet foods, aquatic food, milk foods, and foods for laboratory animals. Extrusion offers a way to cook, shape, and pasteurize these products. It permits better utilization of available cereal grains, as well as vegetable and animal proteins, in manufacturing cost-effective and nutritionally balanced diets with unique and improved characteristics.

Even though the applications of extrusion technology in food processing and manufacturing have broadened over the last two decades, the science and fundamental understanding of the process have always lagged behind the technology. The complex, multi-component, irreversibly reactive nature of food materials, combined with complex screw and die geometries, make complete deterministic analyses of extrusion processes extremely challenging. Issues related to the predictability and control of the flavor and texture of extruded products still need more research and development. In the area of nutraceuticals (health-promoting compounds found mainly in plants), extrusion might offer a way to improve their extractability, enhance their bioavailability, and deliver them effectively.

Extrusion Equipment

Extruders come in several designs, dependent upon their application. Some extruders are designed simply to convey the raw materials, while others are designed to mix and knead them; most, however, are designed to impart mechanical and thermal energy to the raw materials to bring about desired physico-chemical changes. Extruders can be broadly categorized on the basis of the number of screws. The most commonly used extruders are single- and twin-screw. Extruders with more than two screws have been used in the plastics industry but not in food processing.

Single-Screw Extruders

Single-screw extruders contain a single rotating screw in a metal barrel, and come in varying patterns. The main design features are shown in Figure. The most commonly used single-screws have a constant pitch. Single-screws usually consist of three sections: (1) feed, (2) transition or compression, and (3) metering, as shown in figure. The raw materials are fed in a granular form at the hopper located in the feed section. The rotating action of the screw conveys the material to the transition section. In the transition section, the screw channel becomes shallower and the material is compacted. A major portion of mechanical energy is dissipated in this section, which results in a rise in temperature of the material. Starch becomes gelatinized, and the material becomes more cohesive. It is transported further by the metering section and pushed through the die opening. The barrels of single-screw extruders usually have helical or axial grooves on the inner surfaces. This helps to convey and mix the material more effectively.

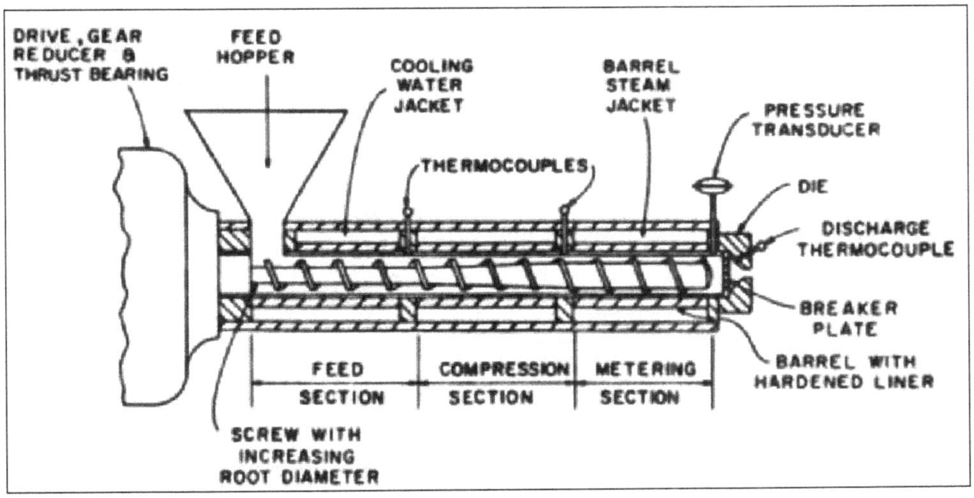

Schematic diagram of a typical single-screw food extruder.

Single-screw extruders are usually characterized by their length to diameter (L/D) ratio and their compression ratio, which is the ratio of the maximum channel depth to the minimum channel depth. The most commonly used compression ratio is 3 to 1. The throughput (mass flow rate) capacity of a single-screw extruder is linked to screw speed, screw geometries, and material characteristics.

Single-screw extruders can be further characterized depending upon the degree of shear they apply. Table shows how single-screw extruders are categorized according to their shearing ability and the associated operating variables. The amount of shear applied results in different product properties. High shear levels cause disruption of starch and protein molecules, affecting their functionality, such as solubility, viscosity, and water holding capacity. Today, approximately 70 percent of the extruders in the food industry are of single-screw type.

Table: Operating characteristics of single-screw extruders.

Operating variable	Low-shear forming extruder	Moderate-shear cooking extruder	High-shear cooking extruder
Feed moisture, % wb	25–35	20–30	12–20
Maximum product temperature, °C	50–80	125–175	150–200
Length to diameter ratio	5–8	10–20	4–12
Diameter to channel height ratio	3–4.5	5–10	7–12
Compression ratio	1:1	2–3:1	3–5:1
Screw speed, rpm	3–4	10–25	30–45
Shear rate, s^{-1}	5–10	20–100	100-180
Net mechanical energy input, $\frac{kW-hr}{kg}$	0.03–0.04	0.02–0.04	0.10–0.14
Heat transfer through barrel jackets, $\frac{kW-hr}{kg}$	−0.01	0.0–0.03	−0.03–0.0
Steam injection, $\frac{kW-hr}{kg}$	0.0	0.0–0.04	0.0
Net energy input to product, $\frac{kW-hr}{kg}$	0.02–0.03	0.02–0.11	0.07–0.14
Product types	Macaroni, RTE cereal pellets, 2nd generation snacks	Soft moist pet foods, pregelatinized starch, drink and soup bases, textured plant protein, RTE breakfast cereals	Puffed starch, dry pet foods, modified starch

Food Drying and Dehydration

The preservation of foods by drying is the time-honored and most common method used by humans and the food processing industry. Dehydration of food is one of the most important achievements in human history, making humans less dependent upon a daily food supply even under adverse environmental conditions. Though in earlier times drying was dependent on the sun, nowadays many types of sophisticated equipment and methods are used to dehydrate foods. During the past few decades, considerable efforts have been made to understand some of the chemical and biochemical changes that occur during dehydration, and develop methods for preventing undesirable quality losses.

State of Water in Foods

The terms dried and dehydrated are not synonymous. Dehydrated foods are those with no more than 2.5% water (dry basis), while the term dried foods applies to any food product with more than 2.5% water (dry basis). The concept of bound water and free water has been developed from drying principles, and it is important for dried products – for its stability during processing and storage. A product containing no water is termed as bone-dry. Water exists in foods in different forms or states. In foods, water having properties different from those of pure water can be defined as bound water. However, the fraction of bound water depends on its definition and the measurement technique used. The binding energy of different states of bound water affects the drying process, since it requires more energy to remove bound water than free water.

End Point of Drying

Equilibrium in drying system is the ultimate endpoint for the process. Water activity is commonly used to estimate the equilibrium point in the case of thermal and osmotic drying processes. In mechanical dewatering, the magnitude of the applied force and rheological properties of the foods affect the equilibrium point. Generally meat, fish, and dairy products are dehydrated to moisture content of 3% or less, vegetable products usually to 5%, and cereal products frequently to as much as 12%. A maximum moisture level is usually established for each dried product separately, based on the desired quality after drying and during storage. Different attributes of quality can be targeted; thus, the endpoint should be determined from all aspects, such as safety first and then consumer acceptance.

Heating Methods in Drying

Heating air using either an electric heater or flue gas is the conventional heating method used for drying foods. In this case, heat transfer from the gas to the product occurs mainly through convection. The heating method is another important aspect of drying, in terms of quality as well as energy cost. Microwave, infrared, radio frequency, refractance window, and dielectric heating use electromagnetic wavelength spectrum as a form of energy, which interacts with the materials, thus generating heat and increasing the drying rate dramatically. Dielectric drying uses frequencies in the range of 1–100 MHz, whereas microwave drying uses frequencies in the range of 300–300,000 MHz. Microwave heating is rapid, more uniform in the case of liquids, and more energy efficient than the hot-air method. Applying microwave energy under vacuum affords the advantages of both vacuum drying and microwave drying, thereby providing improved energy efficiency and product quality. The energy can be applied in pulsed or continuous mode; however, pulsed microwave drying is more efficient than continuous drying. The use of electro-technology in drying is gaining priority in the food industry to improve drying efficiency as well as quality.

Drying Methods

Drying processes can be broadly classified, based on the water-removing method applied, as (a) thermal drying, (b) osmotic dehydration, and (c) mechanical dewatering. In thermal drying, a gaseous or void medium is used to remove water from the material. Thus, thermal drying can be divided further into three types: (a) air drying, (b) low air environment drying, and (c) modified atmosphere drying. In osmotic dehydration, a solvent or solution is applied to remove water, whereas in mechanical dewatering physical force is used. Consideration should be given to many factors before selecting a drying process. These factors are (a) the type of product to be dried, (b) desired properties of the finished product, (c) allowable temperature tolerance, (d) the product's susceptibility to heat, (e) pre-treatments required, (f) capital and processing costs, and (g) environmental factors. There is no one best technique of drying that is applicable for all products.

Thermal Drying

Thermal drying is one of the most widely used methods of drying foods. In this process, heat is mainly used to remove water from the foods. The mechanisms of moisture transfer depend mainly on the types or physicochemical state of food materials and the drying process. Food materials can be classified as (a) homogeneous gels, (b) porous materials with interconnecting pores or capillaries, and (c) materials having an outer skin that is the main barrier to moisture flow. The type or structure of foods always plays an important role in the drying process.

Drying Curve

Drying curve usually plots the drying rate versus drying time or moisture contents. Three major stages of drying can be observed in the drying curve:

- Transient early stage, during which the product is heating up (transient period).
- Constant or first period, in which moisture is comparatively easy to remove (constant rate period).
- Falling or second period, in which moisture is bound or held within the solid matrix (falling rate period).

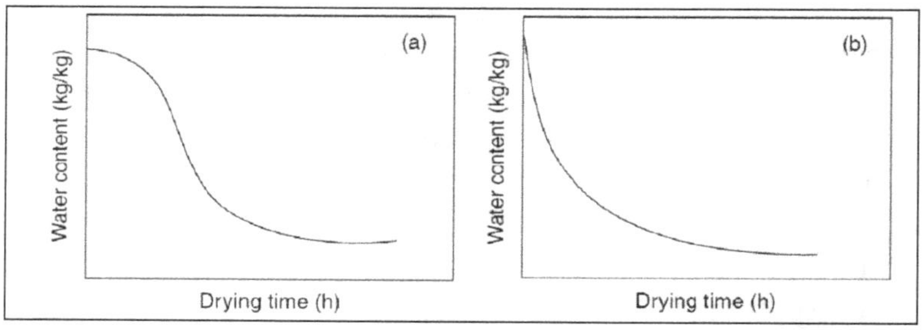

Typical drying curves (water content versus drying time) – (a): With a lag period, (b) Without lag period.

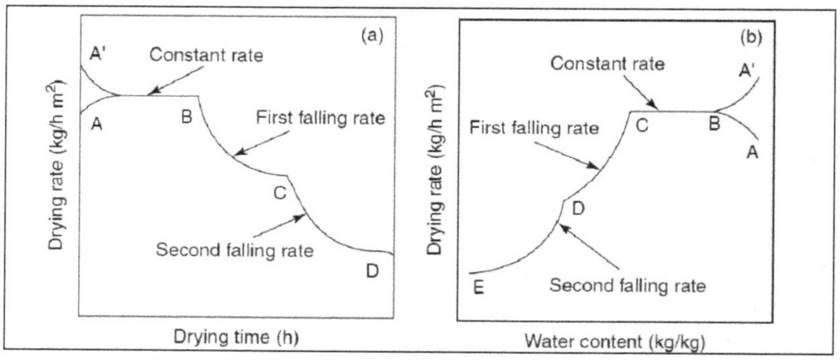

Typical drying curves: (a) Drying rate versus drying time, (b) Drying rate versus water content.

Typical drying rate curves are shown in Figures. The moisture content at which the change from the first to the second period occurs is known as the critical moisture content. Typically, two falling rate periods are observed for both hygroscopic and nonhygroscopic solids. The first falling rate period is postulated to depend on both internal and external mass transfer rates; while the second period, during which drying is much slower, is postulated to depend entirely on internal mass transfer resistance. The slower rate may be due to the solid–water interaction or glass–rubber transition. The drying behaviors of food materials depend on the porosity, homogeneity, and hygroscopic properties. The immediate entrance into the falling rate is characteristic of hygroscopic food materials.

Air-Drying Methods

In the case of air drying, atmosphere is used as the drying medium and heat as different modes could be applied in the process.

Sun Drying

Earlier, only sun drying was used for drying. In this process, foods are directly exposed to the sun by placing them on the land or left hanging in the air. The main disadvantages of this type of drying are (i) contaminations from the environment, (ii) product losses and contaminations by insects and birds, (iii) floor space requirements, (iv) difficulty in controlling the process, and (v) bad odor. When the climate is not particularly suitable for air drying or better quality is desired, mechanical air drying is mainly used. However, sun drying is the cheapest method of drying foods. Nowadays, solar and mechanical air drying is widely used commercially.

Solar Drying

Solar drying is an extension of sun drying that uses radiation energy from the sun. Solar drying is a non-polluting process and uses renewable energy. Moreover, solar energy is an abundant energy source that cannot be monopolized. However, solar drying has several drawbacks that limit its use in large scale production. These are the need for large areas of space and for high labor inputs, difficulty in controlling the rate of drying, and insect infestation and microbial contamination.

In-Store Drying

In-store drying can also be called low-temperature in-bin drying. It may be used when grains are stored until milled or sold. Weather conditions in tropical climates are less favorable for in-store drying, due to high ambient temperatures and relative humidity values. Two-stage drying can produce good quality by preventing discoloration of high-moisture grains and reduced cracking of skin dry kernels.

Convection Air Drying

Cabinet- and bed-type dryers (i.e., kiln, tray, truck tray, rotary flow conveyor, and tunnel) fall into the first generation. This is the simplest drying technique, which takes place in an enclosed and heated chamber. The drying medium, hot air, is allowed to pass over the product, which has been placed in open trays. Convection drying is often a continuous process and is mostly used for products that are relatively low in value. Air drying is usually accomplished by passing air at regulated temperature and humidity over or through the food in a dryer. Factors that affect the rate of drying are temperature, humidity, air velocity and distribution pattern, air exchange, product geometry and characteristics, and thickness. The sample is usually placed on mesh trays in one layer or in bulk on a bed or hung from a string for better air circulation over the product.

Air circulation can be horizontal or vertical to the layer or bed. The structure and composition, such as fat content, of a product affects the drying rate. In general, hotter the air temperature, the faster is the drying rate; and similarly, the higher is the velocity, the higher is the drying rate; the lower is the air humidity, the higher is the drying rate. The relative humidity (a measure of dryness) is lower when air temperature is raised. A dryer must expel air to get rid of moisture, thereby allowing new, lower humidity air to enter the system. However, this process causes heat loss from the dryer. In many cases, two or multistage drying with different conditions could be used, for example, initial drying at 90 °C and then the second or final stage at 60 °C.

Explosive Puff Drying

Explosive puff drying uses a combination of high temperature and high pressure, and a sudden release of the pressure (explosion) to flush superheated water out of a product. This method gives a product of good rehydrability. However, the high heat can degrade food quality, and the explosion puffing may compromise product integrity.

Spray Drying

Spray drying is used to remove water from a free-flowing liquid mixture, thus transforming it into a powder form. The fluid to be dried is first atomized by pumping it through either a nozzle or a rotary atomizer, thus forming small droplets with large surface areas. The droplets immediately come into contact with a hot drying gas, usually air. The liquid is very rapidly evaporated, thus minimizing contact time and heat damage. Disadvantages include the size of the equipment required to achieve drying is very large and very oily materials might require special preparation to remove excessive levels of fat before atomization. Ultrasonication in the chamber can be used instead of complex atomization to produce small-diameter droplets in spray drying.

Fluidized Bed Drying

This technique involves the movement of particulate matter in an upward-flowing gas stream, usually hot air. Fluidization mobilizes the solid particulates, thus creating turbulences on the solid surfaces, which increases the drying rate. The hot gas is introduced at the bottom of a preloaded cylindrical bed and exits at the top. In some cases, a vibratory mechanism is used to increase the contact of the product with the hot gas. Fluidized bed drying is usually carried out as a batch process and requires relatively small, uniform, and discrete particles that can be readily fluidized. The main advantages of fluidized bed drying are uniform temperature and high drying rates, thus less thermal damage. A rotating chamber is also used with the fluidized bed, thus increasing centrifugal force to further increase the drying rate and mixing. The use of a solid carrier, such as sea sand, and wheat bran could prevent the biomaterial from deterioration due to thermal shock.

Spouted Bed Drying

In a spouted bed dryer, a jet of heated gas enters the chamber at the center of a conical base. The food particles are rapidly dispersed in the gas, and drying occurs in an operation similar to flash drying. This works very well with large pieces that cannot be dried in a fluidized bed dryer.

Ball Drying

In this method, the material to be dried is added at the top of the drying chamber through a screw conveyor. The material within the drying chamber comes into direct contact with heated balls made from ceramic or other heat-conductive material. Drying occurs primarily by conduction. Hot air is passed through the bottom side of the chamber. When the product arrives at the bottom of the chamber, it is separated from the balls and collected.

Rotary Drum Drying

Rotary drum dryers are cylindrical shells 1–5 m in diameter, 10–40 m in length, and rotating at 1–8 rpm with a circumferential speed of approximately 0.2–0.4 m/s. These conditions depend on the product types to be dried. The dryers are designed to operate at a nearly horizontal position, inclined only by 2°–6° to maintain the axial advance of solids, which are fed from the upper end of the dryer body.

Drum Drying

This technique removes water from a slurry, paste, or fluid that has been placed on the surface of a heated drum. The dryer may comprise either a single or a double drum. Drum drying is typically a continuous operation, and care must be taken to ensure that the product that is to be dried adheres well to the drying surface; in some cases, it may be necessary to modify the liquid product by using additives to change its surface tension or viscosity.

Low Air Environment Drying

Vacuum Drying

Vacuum drying of food involves subjecting the food to a low pressure and a heating source. The vacuum allows the water to vaporize at a lower temperature than at atmospheric conditions, thus foods can be dried without exposure to high temperature. In addition, the low level of oxygen in the atmosphere diminishes oxidation reactions during drying. In general, color, texture, and flavor of vacuum-dried products are improved compared with air-dried products. In some cases, the product is comparable to the quality of freeze-dried foods.

Freeze Drying

In freeze drying, frozen material is subjected to a pressure below the triple point (at 0 °C, pressure: 610 Pa) and heated to cause ice sublimation to vapor. A schematic diagram of the different states of water with triple point is shown in Figure. This method is usually used for high-quality dried products, which contain heat-sensitive components such as vitamins, antibiotics, and microbial culture. The virtual absence of air and low temperature prevents deterioration due to oxidation or chemical modification of the product. It also gives very porous products, which results in high rehydration rates. However, freeze drying is a slow and expensive process. A long processing time requires additional energy to run the compressor and refrigeration units, which makes the process very expensive for commercial use. Thus, it is mainly used for high-value products.

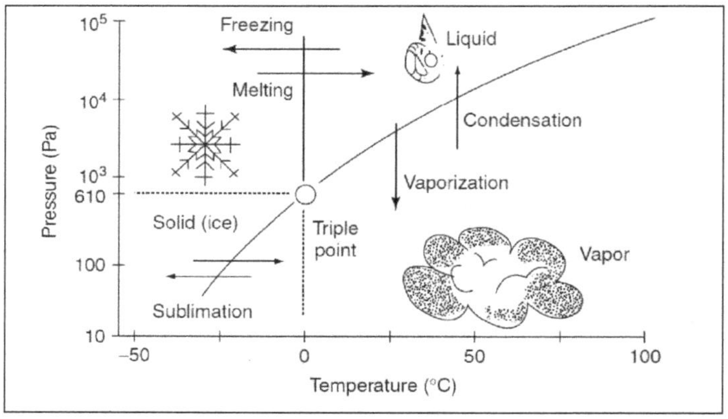

Schematic diagram of the different states of water showing triple point.

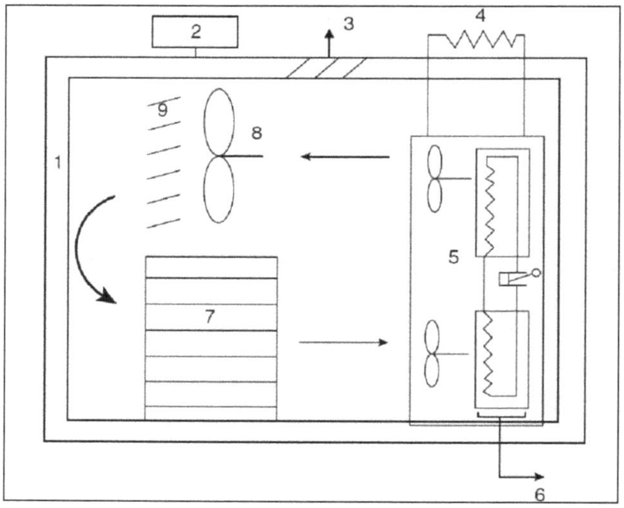

Schematic diagram of operation of a typical heat pump dryer. (1) Vapor – scaled & insulated structure; (2) Humidifier; (3) Over heat vent; (4) External condenser; (5) Heat pump dehumidifier; (6) Condensate; (7) Product tray; (8) Primary sir circulation fan & (9) Air distributor.

Heat Pump Drying

The heat pump dryer is a further extension of the conventional convection air dryer with an inbuilt refrigeration system. Dry heated air is supplied continuously to the product to pick up moisture. This humid air passes through the evaporator of the heat pump where it condenses, giving up its latent heat of vaporization to the refrigerant in the evaporator. This heat is used to reheat the cool dry air passing over the hot condenser of the heat pump. Thus, the latent heat recovered in the process is released at the condenser of the refrigeration circuit and used to reheat the air within the dryer.

The use of the heat pump dryer offers several advantages over conventional hot air dryers for drying food products, including higher energy efficiency, better product quality, the ability to operate independent of outside ambient weather conditions, and zero environmental impact. In addition, the condensate can be recovered and disposed of in an appropriate manner, and there is also the potential to recover valuable volatile components from the condensate. One of the main reasons of quality improvements in heat pump dried products is due to its ability to operate at low temperatures. If a heat pump dryer is used at low temperatures (10 °C–60 °C) for highly perishable food products, adequate precautions need to be taken. There is also potential to use heat pump drying with modified atmospheres to obtain better quality products.

Superheated Steam Drying

Superheated steam is used as a drying medium. The main advantages of this type of drying are that it can provide an oxygen-free medium for drying, and process steam available in the industry can be used without any capital cost. An oxygen-free medium has the potential to provide high-quality food products; however, it is important to generate more information regarding quality improvement and processing efficiency.

Impingement Drying

Impingement drying is an old technology that has only recently been applied to food products. An impingement dryer consists of a single gas jet (air or superheated steam) or an array of such jets, impinging normally on a surface. There are a great variety of nozzles that can be used, and selection of the nozzle geometry and multinozzle configuration have important relevance on the initial and operating costs, and product quality. Some characteristics of impingement drying include rapid drying, popular for convection drying, and the large variety of nozzles available (multizones). Typically, the temperature and jet velocity in impingement drying may range from 100 °C to 350 °C and from 10 to 100 m/s, respectively.

Smoking

Smoking foods is one of the most ancient food preservation processes, and in some communities one of the most important. The use of wood smoke to preserve foods is nearly as old as open-air drying. Although not primarily used to reduce the moisture content of food, the heat associated with the generation of smoke also causes a drying effect. Smoking has been mainly used with meat and fish. The main purposes of smoking are it imparts desirable flavors and colors to the foods, and some of the compounds formed during smoking have a preservative effect (bactericidal and antioxidant) due to the presence of a number of compounds. In many cases, smoking is considered as a pre-treatment rather than a drying process. It was found that smoke is effective in preventing lipid oxidation in meat and fish products.

Modified Atmosphere Drying

This is a new concept of drying foods using heat pump dryers, which uses modified atmospheres such as nitrogen and carbon dioxide, for better quality and preservation of constituents of foods, prone to oxidation. Technologies to create the modified atmosphere drying are now evolving. Modified atmosphere heat pump dehumidifier (MAHPD) drying is a relatively new development. The fact that heat pump dehumidifier (HPD) drying is conducted in an enclosed, insulated chamber is made use of in the development of the MAHPD drying system. The air in the dehumidifier chamber is replaced with an inert atmosphere such as nitrogen, carbon dioxide, or their mixtures. Replacement of the air inside the chamber is easily carried out by exhausting the chamber using a vacuum pump and then breaking the vacuum using an inert gas. Vacuum exhaustion is a more cost-effective way to replace air than by direct purging with the specific inert gas.

Replacement of air with carbon dioxide or nitrogen by purging requires over 50 volumes to achieve an oxygen level of less than 0.5%. Schematic diagram of the MAHPD drying system is shown in figure. This consists of a sealed drying vessel connected to the heat exchanger unit. The drying vessel has provision for introduction of nitrogen or for evacuation through a valve connection. The MAHPD system shown also has provisions for introducing microwave energy for heating the product, through a slotted waveguide running down the wall of the chamber parallel to the axis of the chamber. A PLC control panel connected to a remote PC and monitor controls the whole system. The product is carried on microwave-transparent plastic trays stacked vertically on a rotating platform, which is mounted on a load cell, so that weight loss can be monitored and recorded on the PC.

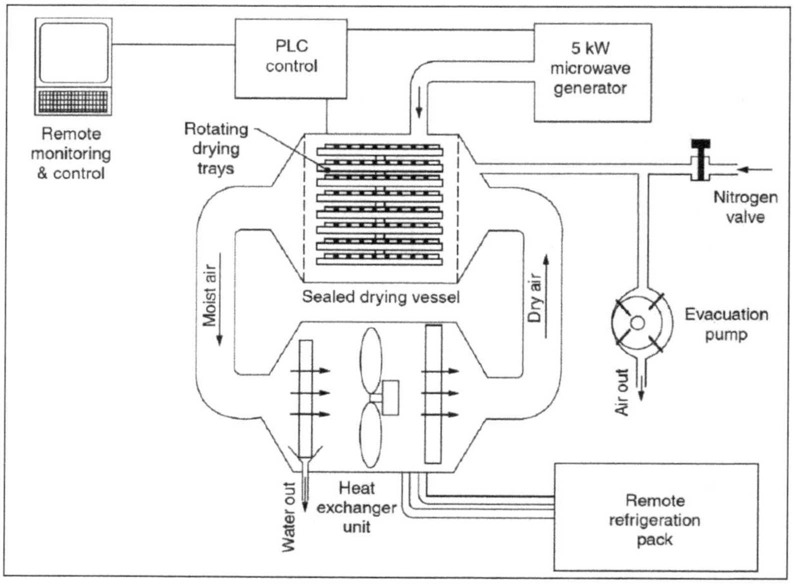

Schematic diagram of MAHPD drying system.

Pre-treatments

Pre-treatments are common in most of the drying processes in order to improve product quality, storage stability, and process efficiency. In recent years, an improvement in quality retention of the dried products by altering processing strategy and pre-treatment has gained much attention.

1. Blanching: Blanching is a process of preheating the product by immersion in water or steam. The main purpose of blanching is to inactivate the naturally occurring enzymes present in foods, since enzymes are responsible for off-flavor development, discoloration or browning, deterioration of nutritional quality, and textural changes in food materials. Other advantages are that it removes air bubbles from vegetable surfaces and from intercellular spaces, reduces the initial microbial load, cleans raw food materials initially, facilitating preliminary operations such as peeling and dicing, and improving color, texture and flavor under optimum conditions. Blanching may have disadvantages, for example, it may change the texture, color, and flavor because of the heating process; it increases the loss of soluble solids, such as vitamins, especially in the case of water blanching; it may change the chemical and physical states of the nutrients and vitamins; and it has

adverse environmental impacts, such as large water and energy usage, and problems of effluent disposal.

2. Sulfur Dioxide Treatment: Sulfur dioxide preserves the texture, flavor, vitamin content, and color that make food attractive to the consumer. Sulfur dioxide treatment is used widely in the food industry to reduce the fruit-darkening rate during drying and storage, and preserves ascorbic acid and carotene. Sulfur dioxide taken up by the foods displaces air from the tissue in plant materials, softens cell walls so that drying occurs more easily, destroys enzymes that cause darkening of cut surfaces, shows fungicidal and insecticidal properties, and enhances the bright attractive color of dried fruits. Permitted levels of sulfur dioxide and other additives (solutes) in dried foods vary from country to country.

3. Salting or Curing: Salting or curing is a natural type of osmotic dehydration. Curing was originally developed to preserve certain foods by the addition of sodium chloride. In the food industry, the application of curing is related only to certain meat, fish, and cheese products. Today sodium chloride, and sodium and potassium nitrite (or nitrate) are considered as curing salts. Salting is one of the most common pre-treatments used for fish products. It converts fresh fish into shelf-stable products by reducing the moisture content and acting as a preservative. In combination with drying, these processes contribute to the development of characteristic sensory qualities in the products, which influence their utilization as food.

4. Freezing Pre-treatment: Freezing treatment affects the drying process. The rehydration rate of air- and vacuum-dried fruits and vegetables subjected to freezing treatment increased to a level comparable with that of freeze-dried products. It is also noticed that the longer the duration of freezing, the better the rehydration kinetics of dried products. This is due to the formation of large ice crystals by slow freezing.

5. Cooking: Cooking at different pressure levels before drying can destroy microorganisms and affect the physicochemical properties of dried products. The bacterial load on the final product can thus be reduced considerably, and the cooked product can be minced and spread evenly on drying trays with much less trouble than the raw material. Precooking is usually used for rice, beef, fish, and beans. Formation of superficial pellicle (case-hardening) may be avoided by precooking, which considerably retards drying. It is clear that the more severe the initial conditions of cooking, the more stable is the subsequently dehydrated product. Generally, cooked vegetable, meat, or fish is dried more easily than their fresh counterparts, provided that the cooking does not cause excessive shrinkage or toughening. Cooking also results in a decrease in water-holding capacity of meat products.

6. Other Dipping Pre-treatments: Dipping treatment with chemicals is also used in addition to blanching or sulfite treatment. The dipping treatment is a process of immersion of foods in a solution containing additives. Usually, the concentration level is below 5% and the dipping time below 5 minutes, whereas osmotic dehydration is carried out at higher concentrations and for long processing times. The main purpose of the dipping treatment is to improve the drying characteristics and quality. Certain chemicals are used to enhance the rate of dehydration. Among these compounds, methyl and ethyl oleate, or olive oil are the most common. Table shows the chemicals used for dipping treatment.

Table: Chemicals used for dipping treatment.

Type Chemicals	Compounds
Esters	Methyl oleate, ethyl oleate, butyl oleate
Salts	Potassium carbonate, sodium carbonate, sodium chloride, potassium sorbate, sodium polymetaphosphate
Organic acids	Oleic acid, steric acid, caprillic acid, tartaric acid, oleanolic acid
Oils	Olive oil
Alkali	Sodium hydroxide
Wetting agents	Pectin, tween, nacconol
Others	Sugar, liquid pectin
Surfactants	
Nonionic	Monoglycerides, diglycerides, alkylated aryl polyester alcohol, polyoxyethylene sorbitan monostearate, sorbitan monostearate, Dsorbitol, polyoxyethylene
Anionic	Sodium oleate, steric acid, sorbitan heptadecanyl sulfate, dimethyl-benzyl-octyl ammoniumchloride

Quality Changes during Drying

1. Selection of Variety: Optimum freshness plays an important role in determining the quality and stability of dried foods; fresher the raw material, more stable and better is the quality of the product. Suitable varieties of produce with the desired maturity should be used to achieve a product that is best in quality. The quality characteristics of dried foods can be grouped as microbial, chemical, physical, and nutritional.

2. Microflora in Dried Foods: Multiplication of microorganisms should not occur in properly processed dehydrated foods, but they are not immune to other types of food spoilage. If dried foods are safe in terms of pathogenic microbial count and toxic or chemical compounds, then acceptance depends on the flavor or aroma, color, appetizing appearance, texture, taste, and nutritional value of the product. Microbial standards are usually based on the total number of indicator organisms or number of pathogens. The microbial load and its changes during drying and storage are important for establishing a standard that will ensure food safety.

Table: Quality Characteristics of Dried Foods.

Microbial	Chemical	Physical	Nutritional
Pathogens	Browning	Rehydration	Vitamin loss
Spoiling	Oxidation	Solubility	Protein loss
Toxin	Color loss	Texture	Functionality loss
	Aroma development	Aroma loss	Fatty acid loss
	Removal of undesired components	Porosity	
		Shrinkage	
		Pores' characteristics	
		Crust formation	
		Structure	

3. Browning Reactions: Browning reactions change color, decrease nutritional value and solubility, create off-flavors and induce textural changes. Browning reactions can be classified as enzymatic or non-enzymatic, with the latter being more serious as far as the drying process is concerned. The two major types of non-enzymatic browning are caramelization and Maillard browning. In addition to moisture level, temperature, pH, and composition are the other parameters that affect the rate of non-enzymatic browning. The rate of browning is most rapid in the intermediate moisture range and decreases at very low and very high moistures. Browning tends to occur primarily at the center of the drying period. This may be due to the migration of soluble constituents toward the center. Browning is also more severe near the end of the drying period, when the moisture level of the sample is low and less evaporative cooling is taking place that causes the product temperature to rise.

4. Lipid Oxidation: Dehydrated foods containing fats are prone to develop rancidity after a period, particularly if the water content is reduced too much. Fish oils or fats are more unsaturated than beef or butter, and they are usually classified as drying oils because they contain considerable proportions of highly unsaturated acids. The behavior of drying oils toward atmospheric oxygen is well known, and oxidation is a serious problem for commercial drying of fatty fish and seafood. The flesh of some fatty fish, such as herrings, contains a fat pro-oxidant that is not wholly inactivated by heat. Lipid oxidation is responsible for rancidity, development of off-flavors, and the loss of fatsoluble vitamins and pigments in many foods, especially in dehydrated foods. Factors that affect oxidation rate include moisture content, type of substrate (fatty acid), extent of reaction, oxygen content, temperature, presence of metals and natural antioxidants, enzyme activity, UV light, protein content, free amino acid content and other chemical reactions. Moisture content plays a big part in the rate of oxidation. At water activities around the monolayer (aw 0.3), resistance to oxidation is greatest. The elimination of oxygen from foods can reduce oxidation, but the oxygen concentration must be very low.

5. Changes in Proteins: The protein matrix in muscle has a marked effect upon its functionality and properties. The non-fatty part of fish is very susceptible to changes caused by the high temperature of initial cooking, as well as drying and storage. Every process involved in the conversion of muscle to meat alters the characteristics of the structural elements. Heating is believed to cause the denaturation of the muscle proteins even below 60 °C but not enough to greatly shear resistance.

6. Structural Changes: Structural changes in food during drying are usually studied by microscopy. Microscopy provides a good tool to study this type of phenomenon as well as other types of physical and chemical changes during the drying of food materials. Shrinkage occurs first at the surface and then gradually moves to the bottom as the drying time increases. The cell walls become elongated. As drying proceeds at high temperature, cracks are formed in the inner structure. Rehydration is maximized when cellular and structural disruption such as shrinkage is minimized.

7. Case Hardening or Crust Formation: During drying, the concentration of moisture in the outer layers of foods is less than in the interior, since the outer layers necessarily lose moisture before the interior. This surface shrinkage causes checking, cracking, and warping. This type of shrinkage causes moisture gradient and resistance near the surface. In extreme cases, shrinkage and drop in diffusivity may combine to yield a skin practically impervious to moisture, which encloses the

volume of the material so that the interior moisture cannot be removed. This is called case hardening. In food processing, case hardening is also commonly known as crust formation. The extent of crust formation can be reduced by maintaining flattening moisture gradients in the solid, which is a function of drying rate. The faster the drying rate, the thinner the crust. Crust (or shell) formation may be either desirable or undesirable in dried food products.

8. Shrinkage or Collapse and Pore Formation: Two types of shrinkage are usually observed in the case of food materials isotropic and anisotropic shrinkage. Isotropic shrinkage can be described as the uniform shrinkage in all geometric dimensions of the materials. Anisotropic shrinkage is described as the non-uniform shrinkage in the different geometric dimensions. Shrinkage is an important phenomenon impacting dried food product quality by reducing product wettability, changing product texture, and decreasing product absorbency. Depending on the end use, crust and pore formation may be desirable or undesirable. If a long bowl life is required for a cereal product, a crust product that prevents moisture reabsorption may be preferred. If a product (such as dried vegetables in instant noodles) with good rehydration capacity is required, high porosity with no crust is required.

9. Stress Development and Cracking or Breakage: During air drying, stresses are formed due to non-uniform shrinkage resulting from non-uniform moisture and temperature distributions. This may lead to stress crack formation when stresses exceed a critical level. Crack formation is a complex process influenced interactively by heat and moisture transfer, physical properties, and operational conditions. The relative humidity of air and temperature are the most influential parameters that need to be controlled to eliminate the formation of cracks. Checking and breakage of dried foods has two undesirable consequences––loss of valuable product and loss of consumer satisfaction.

10. Rehydration: Rehydration is a process of moistening dry material. It is mostly done by applying an abundant amount of water. In most cases, dried foods are soaked in water before cooking or consumption, thus rehydration is one of the important quality criteria. In practice, most of the changes during drying are irreversible and rehydration cannot be considered simply as a process reversible to dehydration. In general, absorption of water is fast at the beginning and thereafter slows down. This rapid moisture uptake is due to surface and capillary suction.

11. Volatile Development or Retention: In addition to physical changes, drying generates flavor or releases flavor from the foods. Drying changes the composition of volatiles by evaporating most volatiles and forming new volatile odor compounds by chemical reactions. Such changes in volatiles might affect the aroma of fresh foods after drying.

12. Solubility: Many factors affect solubility, including processing conditions, storage conditions, composition, pH, density, and particle size. It was found that an increase in drying temperature is accompanied by increasing protein denaturation, which decreases solubility. Thus, more protein is denatured and solubility decreased.

13. Caking and Stickiness: Caking and stickiness of powders, desirable or undesirable, occur in dried products. Caking is desirable for tablet formation and undesirable when a dry free-flowing material is required. To reduce caking during drying, a logical option is to dry rapidly so that the moisture content drops to a level where caking is inhibited. The rapid drying will form a crust,

which may be undesirable, thus product optimization or solutes in product formulation may be considered.

14. Texture: Factors that affect texture include moisture content, composition, variety or species, pH, product history (maturation or age), and sample dimensions. Texture is also dependent on the method of dehydration and pre-treatments.

15. Vitamins Retention: In general, losses of B vitamins are usually less than 10% in dried foods. Dried foods do not greatly contribute to dietary requirements for thiamin, folic acid, and vitamin B-6. Vitamin C is largely destroyed during drying due to heating. From non-fatty vegetables, such as cabbage, as much water as possible should be removed, because this helps to conserve ascorbic acid. The loss of vitamin A and ascorbic acid in dried products could be avoided in the absence of oxygen. Even though most amino acids are fairly resistant to heating–drying, lysine is quite heat labile and likely to be borderline or low in the diet of humans and especially so in developing countries where high quality animal proteins are scarce and expensive.

16. Color Retention or development: High temperature and long drying time degrade a product's original color. Color in foods can be preserved by minimal heat exposure or applying high temperature and short time with pH adjustment. Water activity is one of the important factors degrading chlorophyll. Another cause of color degradation may be due to enzymatic browning causing rapid darkening, mainly of the leafy portions. The formation of dark pigments via enzymatic browning is initiated by the enzyme polyphenol oxidase (PPO). Another reason for discoloration is photo oxidation of pigments, caused by light in combination with oxygen.

Roasting and Smoking of Foods

Food processing techniques dates back to the prehistoric ages. Cooking is evolved as one of the widely practiced food preservation method. Cooking is a type of heating process in which the primary objective is to produce a more palatable food. During cooking heat is transferred to the food either by conduction, convection, radiation or by the energy of microwaves-electronic heat transfer. In general, cooking takes place either by moist or dry heat as mentioned in the following table. Water or steam and air or fat or combinations of these are used as cooking media. Moist heating (cooking) involves water and steam. Air or fat are most commonly used in dry heat.

The word cooking is a broad term embodying at least six forms of heating, including baking, broiling, roasting, boiling, frying, and stewing. The method of applying heat energy and the duration differs somewhat for each of these processes. Baking, broiling, and roasting usually require dry heat at relatively high temperatures (greater than 100 °C), boiling and stewing are done by placing the product in boiling water, and frying involves cooking oil and temperatures much greater than 100 °C.

Roasting

Roasting was originally meant turning meat or a bird on a spit in front of a fire. Traditionally recognized roasting methods consist only of baking and cooking over or near an open fire. Grilling is normally not technically a roast, since a grill (gridiron) is used. Barbecuing and Smoking differ from roasting because of the lower temperature and controlled smoke application.

Roasting is generally defined as a dry heat treatment of foods with the intention to generate roast aroma compounds, to develop color, and often to create a crispy texture. These intentional product alterations make the explicit difference between roasting and simple drying. Roasting can enhance flavor through caramelization and Maillard browning on the surface of the food. Roasting is a cooking method that uses indirect, diffused heat (as in an oven) or dry heat, whether in an open flame, oven, or by the application of other heat source. Roasting is applied to a number of foodstuffs. In general any type of food products that has been cooked in this fashion is called a roasted.

Table: Different types of Cooking Techniques.

Type/Mode of Cooking		List of Cooking Techniques
Dry	Conduction	Dry Roasting, Hot Salt frying, Searting
	Conduction	Baking, Roasting (modern), Smoking
	Radiation	Grilling, Roasting (Traditional), Toasting, Rotiserie
Wet	High heat	Blanching, Boiling, Decoction, Parboiling, Shocking
	Low heat	Codding, Infusion, Poaching, Simmering, Steeping, Steawing
	Indirect Heat	Double boiling, Steaming
Fat-based	High heat	Blackening, Deep frying, Pan frying, Shallow frying
	Low heat	Sweating, Gentle frying
Mixed medium	Barbecuing, Braising, Stir frying, Gridding, Indirect grilling	
Device-based	Clay-pot cooking, Earth oven, Masonry oven (Tandoor), Micro oven, Pressure cooker, Pressure frying, Solar oven	
Non-heat	Curing, Fermentation, Pickling, Souring	

Mechanisms of Roasting

Roasting can be used to cook a variety of foodstuffs such as beans, nuts, oilseeds, cereal grains, pulses and most root and bulb vegetables, poultry, meat, fish and seafood. Roasting is a time temperature controlled process that usually involves dehydration, reaction of free amino acids and short-chained peptides with free mono- and disaccharides during non-enzymatic browning, protein denaturation and subsequent changes in texture. Roasting is normally carried out under atmospheric conditions with hot combustion gases and excess air as the primary heating agent though heat may also be provided by contact with hot metal surfaces, solely in more primitive methods, but more generally as a supplement to convection from the hot gases. Some heat may also be supplied by radiation and in some designs even as the primary source by radiative heating of various kinds. Baking and roasting are essentially the same unit operation: they both use heated air to alter the eating quality of foods. The terminology differs in common usage; baking is usually applied to flour-based foods or fruits, and roasting to beans, coffee beans, oil seeds and nuts, meats, nuts and vegetables.

The roasting process can be roughly divided into a drying phase, a roasting phase, where a number of complex chemical reactions take place, and a final cooling phase. In roasting the initial stage of chemical reactions are endothermic, whereas a number of authors state an exothermic final roasting stage. On the basis of calorimetric measurements, the net result of reactions in coffee to become exothermic.

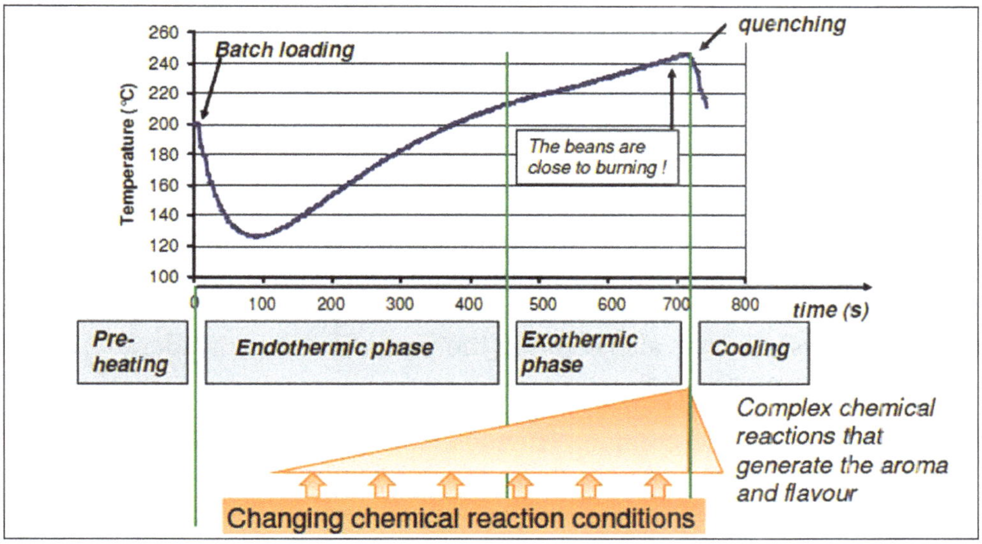

Mechanism of roasting process.

Methods of Roasting

Roasting method differs according to type of food matrix. For those foods which are meant for immediate consumption (meats and fish products) roasting method differs than that of other foods (coffee beans, oil seeds and nuts). There are several plans for roasting: low-temperature roasting, high-temperature roasting, and a combination of both. Each method can be suitable, depending on the food and the tastes of the people. For roasting, the food may be placed on a rack or in a roasting pan to ensure even application of heat, may be rotated on a spit or rotisserie. If a pan is used, the juice can be retained for use in gravy. During oven roasting, hot air circulates around the food, cooking all sides evenly.

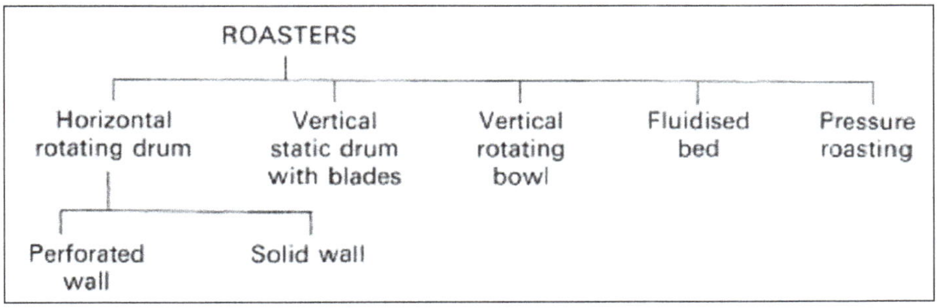

Different types of roasters.

The horizontal rotating drum has been the most usual method of providing contact of hot air in a tumbling coffee bed, but roasters have been designed and are available based on other mechanical principles. Horizontal rotating drums may provide either batch or continuous roasting whereas the vertical drum and bowl are usually batch. Fluidisedbed roasters generally operate in the batch mode, but the latest designs are often continuous.

Physico-Chemical Changes in Roasted Foodstuff

Two important preservative changes occur in food as a result of roasting: destruction or reduction of microorganisms and inactivation of undesirable enzymes. Other desirable changes that may occur during cooking include (1) destruction of potentially hazardous toxins present naturally or through microorganisms; (2) alteration of color, flavor, and texture; and (3) improved digestibility of food components. Undesirable changes also may occur, such as degradation of nutritive components and sensory attributes.

During roasting the beans loses weight, generally between 14 and 20%, depending on the green bean quality, the process conditions and the target degree of roast. A major part of this weight loss is due to dehydration, whereas another substantial part (some 5 to 8% for a medium degree of roast) is caused by a loss of dry matter, primarily as CO_2. The chemical reactions that convert organic matter into gaseous products also result in the formation of a considerable amount of water that is then again lost as water vapor.

Some of the more extensive and complex chemical reactions during roasting affect the carbohydrates of green beans and include Maillard reaction, Strecker degradation, pyrolysis, caramelization, mainly resulting in aroma, flavor and color compounds. Roasting leads to protein denaturation and degradation. Free amino acids, peptides and proteins with free amino groups react with reducing sugars to form glycosylamines and/or aminoaldoses and/or amino-ketones by condensation. Amino acids react with a-dicarbonyls during Strecker degradation and form amino-ketones. On roasting there is a reduction in the amount of citric and malic acid and an increase of many of the other acids, in particular quinic acid and volatile acids. Chlorogenic acids are strongly degraded. The loss is about proportional to the degree of roast and can reach 80% in dark roasted.

Changes in colour and volume during roasting stage with respect to degree of roasting. CTN= Colour Test Neuhaus scale, reflectance colorimetric unit that is used to determine the final roasting point.

As presented in figure, the volume increase is the most obvious macroscopic change of the bean structure during roasting. Bean expansion occurs progressively, but including a "popping phase", leading to considerable decrease of density. It is not quite clear from this statement whether the term “popping phase” applies only to sounds accompanying the expansion, or a phase of instantaneous expansion. Volume increase and density decrease is a function of the degree of roast, but also of the speed of roasting. When the resulting bean volume is correlated with the final roasting temperature; bean expansion is caused by a very rapid pressure build-up, due to rapid formation of water vapor and gas within the bean. A steady and continuous volume increase is positively correlated to weight loss. Extent of darkness is also positively correlated with the degree of roasting.

The roasted whole beans are characterized by the roasting process to which they have been subjected; their so-called degree of roast is reflected in their external colour, the flavour developed, the dry mass loss that has occurred, and the chemical changes in selected components. In simple terms, roast coffee can be described as having a ‘light’, ‘medium’ or ‘dark’ roast. Degree of roast plays a major part in determining the flavour characteristics of extracts subsequently brewed from roasted coffee.

Roasting of Meat and Meat Products

Tenderness of meat changes during roasting due to the transformations in connective tissue and myofibrilar proteins. When heated in the presence of water, collagen dissolves, which results in crushing of meat, while myofibrilar proteins denature, which causes the meat increased hardness and shrinkage both crosswise and along muscle fibres and connective tissue. During heat treatment, meat loses 20-40% of its total initial weight due to fluid leakage with the increasing temperature. Cooking loss is strongly associated with fibres shrinkage and thereby impacts the overall process efficiency and general consumer acceptance of the product. Proper roasting methods and setting of temperature-time-humidity parameters, adequate to these processes, are important in improving tenderness in roasted food.

There is a loss of dry matter, primarily as gaseous carbon dioxide and water (over and above that moisture already present), and other volatile products of the pyrolysis. During roasting, meats and vegetables are frequently basted on the surface with butter or oil to reduce the loss of moisture by evaporation. In recent times, plastic oven bags have become popular for roasts. These cut cooking times and reduce the loss of moisture during roasting, but reduce flavor development from Maillard browning, somewhat more like (boiled or steamed) stew or pot roast. They are particularly popular for turkeys. In general, in either case, the meat is removed from heat before it has finished cooking and left to sit for a few minutes, while the inside cooks further from the residual heat content, a phenomenon known as carry over cooking, also known as “resting” the meat. Following are the three types of roasting employed in meat and meat products.

- A low-temperature oven, 95 °C to 160 °C (200 °F to 325 °F), is best when cooking with large cuts of meat, turkey and whole chickens. This is not technically roasting temperature, but it is called slow-roasting. The benefit of slow-roasting an item is less moisture loss and a tenderer product. More of the collagen that makes meat tough is dissolved in slow cooking. At true roasting temperatures, 200 °C or more, the water inside the muscle is lost at a high rate.

- Cooking at high temperatures is beneficial if the cut is tender enough as in filet mignon or strip loin to be finished cooking before the juices escape. A reason for high temperature roasting is to brown the outside of the food, similar to browning food in a pan before pot roasting or stewing it. Fast cooking gives more variety of flavor, because the outside is brown while the center is much less done.

- The combination method uses high heat just at either the beginning or the end of the cooking process, with most of the cooking at a low temperature. This method produces the golden-brown texture and crust, but maintains more of the moisture than simply cooking at a high temperature, although the product will not be as moist as low temperature cooking the whole time. Searing and then turning down to low is also beneficial when a dark crust and caramelized flavor is desired for the finished product. Note that searing in no way "locks in" moisture: moisture loss is simply a function of heat and time.

Roasting of Coffee Beans

Coffee and cocoa has received much attention because of the high demand for the typical caramalized flavors. After harvesting, cocoa beans are fermented under controlled conditions. The beans are then roasted; that darkens the color and yields a typical characteristics flavor. The fermentation hydrolyzes sucrose to reducing sugars, frees amino acids, and oxidizes some polyphenols. During roasting, many pyrazines and other heterocyclics are formed, but the unique flavor of cocoa is derived from an interaction between aldehydes from the Strecker degradation reaction in roasting stage.

Green coffee beans provide neither the characteristic aroma nor flavour of brewed coffee until they are roasted. Moreover, the roasting process increases the value of coffee beans, by 100- 300% of the raw material. Roasting of coffee beans typically takes place at 200-240 °C for different times depending on the desired characteristics of the final product. Events that take place during roasting are complex, resulting in the destruction of some compounds initially present in green beans and the formation of volatile compounds that are important contributors to the characteristic of coffee's aroma.

Briefly, as temperature increases to about 100 °C, green coffee beans undergo moisture loss from 8-12% in green coffee beans to about 5% in the roasted coffee beans. The smell of the beans changes from herb-like green bean aroma to bread-like, the color turns from green to yellowish, and the structure changes from strength and toughness to more crumbly and brittle. When the internal temperature of beans reaches 100 °C, the color darkened slightly for about 20-60 s due to the vaporization of water. At 160-170 °C, the beans become lighter in color for about 60-100 s. As roasting continues at this temperature, beans begin to turn brown and the caffeol, or oil, locked inside the beans begins to emerge. This process, called pyrolysis, is at the heart of roasting. It is what produces the flavor and aroma of the coffee. Maillard reactions also start to take place, resulting in gradually darkening of the beans. The buildup of water pressure, along with the large amount of gases generated causes the cellulose cell wall to crack, giving rise to the so called "first crack". As heating continues at the roasting temperature (160-170 °C), the coffee becomes darker and more rapid popping of coffee bean occurs ("second crack") as the carbon dioxide (CO_2) buildup exceeds the strength of the cellulosic walls of the bean. Finally, after roasting, the fresh roasted coffee beans are quickly cooled to stop roasting.

The final quality of roasted coffee is influenced by the design of the roasters and timetemperature profiles used. Although heat transfers during roasting can involve conduction, convection, and radiation, convection by far is the most important mode of heat transfer that determines the rate and uniformity of roasting.

Smoking

Smoking is one of the oldest method of food preservation, probably having arisen shortly after the development of cooking with fire. Smoking is the process of cooking, flavoring, or preserving the food by exposing it to the smoke from burning or smoldering plant materials, most often wood.

From the engineering point of view there is no such process as smoking. Smoke is an antimicrobial and antioxidant, used as a condiment and claimed as a preservative, but smoke alone does not give us the so called smoked food products. The basic food process called drying, combined with the effects of salt and smoke particulates results in smoked products. Smoking is defined as the process of penetration of volatiles resulting from thermal destruction of wood into the surface of food products.

Meat, chicken and fish are the most common smoked foods. Cheeses, nuts and seeds, hard-boiled eggs, berries, vegetables, and ingredients used to make beverages such as whisky, Rauchbier, and lapsang souchong tea are also smoked. However, amateurs using ordinary smoke ovens or adapting barbecue grills to the purpose have successfully used the smoking technique to flavour and preserve not only meat, variety meats parts including heart, tongue, and liver. Pork and beef hams, bacon bellies, and sausages are the most common commercially smoked meats.

Types of Smoking

- Cold smoking: Cold smoking can be used as a flavor enhancer for items such as chicken breasts, beef, pork chops, salmon, etc. The item is hung first to develop a pellicle, and then can be cold smoked for just long enough to give some flavor. Some cold smoked foods are baked, grilled, steamed, roasted, or sauted before eating. Smokehouse temperatures for cold smoking are typically done between 20 to 30 °C. In this temperature range, foods take on a smoked flavor, but remain relatively moist. Cold smoking does not cook foods.

- Hot smoking: Hot smoking exposes the foods to smoke and heat in a controlled environment. Like cold smoking, the item is hung first to develop a pellicle, and then smoked. Although foods that have been hot smoked are often reheated or cooked, they are typically safe to eat without further cooking. Hams and ham hocks are fully cooked once they are properly smoked. Hot smoking occurs within the range of 52 to 80 °C. Within this temperature range, foods are fully cooked, moist, and flavorful.

- Smoke-roasting" or "Smoke-baking: Smoke roasting or smoke baking refers to any process that has the attributes of smoking combined with either roasting or baking. This smoking method is sometimes referred to as "barbecuing", "pit baking", or "pit roasting". It may be done in a smoke roaster, closed wood-fired masonry oven or barbecue pit, any smoker that can reach above 121 °C, or in a conventional oven by placing a pan filled with hardwood chips on the floor of the oven so the chips smolder and produce a smoke bath. However, this should only be done in a well-ventilated area to prevent carbon monoxide poisoning.

Mechanism of Smoking

Application of smoke to foodstuffs essentially is a physical process, which is based on such phenomena as diffusion, absorption, dissolution and deposition in force fields. It is accompanied by chemical process where in smoke compounds interact with food components. Te basic fact pertaining to the smoke formation and application is the products of oxidative thermo-destruction of raw materials used for the developing the smoke.

Whether done on a commercial or a home scale, the smoking technique involves hanging the food products or placing it on racks in a chamber designed to contain the smoke. Commercial smokehouses, usually several stories high, often use steam pipes to supplement the heat of a natural sawdust fire. Generally, smokehouse temperatures may go up to 71° C, and smoking periods vary from as short as a few hours to as long as several days, depending on the type of food and its moisture content. In order to shorten the production process, commercial meats are sometimes artificially "smoked" by dipping them in a solution of preservative chemicals or by painting them with such a solution. But because this procedure involves no natural drying action, it has practically no preservative effect.

- Dry Smoking: Dry smoking uses indirect cooking with a low, smoldering wood fire to slowly cook foods while infusing smoke flavor.
- Wet Smoking: Wet smoking, or water smoking, is more commonly employed and uses a pan of water to maintain moisture and tenderness.

Preservation Effect in Smoking

Smoke is an antimicrobial and antioxidant, but smoke alone is insufficient for preserving food in practice, unless combined with another preservation method. The brown color that develops at the surface of protein-based foods during smoking is primarily caused by non-enzymatic browning reactions involving amino groups on food proteins and carbonyl compounds in the smoke. Phenol as one of smoke components act as antioxidant agent, while other components are organic acids, alcohol, carbonyl, hydrocarbon and nitrogen compounds such as nitrooxide, whilst the compounds found at the surface which possibly also penetrate into smoked food are aldehydes, ketones, ester and ether also impart preservative effect.

Smoked salmon and cheese.

Desirable Effects of Smoking on Food

The desirable effects of smoking on foods are flavouring, preservation and coloring, contributed by the smoke material. Wood, the material most widely used for smoke production, consists of many groups of polymers which are the cellulose, lignin and hemicelluloses. The composition of wood smoke is very complex and is known to contain acids, carbonyls, alcohols, phenolics, and other neutral compounds. Many of the fat-soluble compounds present in wood smoke are absorbed into the food surface during natural smoking, and presumably the same is true for water-soluble components such as acids. The typical aroma of smoked food seems to be due mainly to the effect of phenols, carbonyls and acids. The smoke phenols contribute essentially to the typical flavour of smoked food.

Undesirable Effects of Smoking on Food

Undesirable effects are contamination with toxic components of smoke and some destruction of essential amino acids of food proteins which are attributed to certain class of components of smoke and liquid smoke. The main problem is the smoke compounds adhere only to the outer surfaces of the food; smoke does not actually penetrate far into meat or fish. Surface dehydration often occurs in foods that are subjected to natural vaporous smoking. Loss of moisture during smoking should cause soluble constituents to become more concentrated within the water phase of the food material in direct relation to moisture loss. If the smoker is allowed to get hotter than 85 °C, the foods will shrink excessively, buckle, or even split. Smoking at high temperatures also reduces yield, as both moisture and fat are “cooked” away.

Health Implications of Smoked and Roasted Foods

Eating a diet high in smoked, cured, or salted meats has been shown to be a risk factor in stomach cancer. In addition to sugar and salt exposure, smoking can directly create compounds known to have long-term health consequences, namely polycyclic aromatic hydrocarbons, or PAHs, many of which are known or suspected carcinogens.

Food Deterioration and Control

Foods for human consumption can be divided into mainly plant and animal foods. Cereals, sugar products, fruits and vegetables are the main foods from plants. In addition to these are spices and food fungi. Foods from animals are meat, poultry, eggs, fish, seafood, milk and milk products. Sodium chloride is a mineral food, a flavouring material, an essential nutrient and a chemical preservative. Some foods may be fortified with minerals e.g. iron and calcium compounds are added to flour. Some of the coloring and flavoring materials used in foods are synthetic. Vitamins usually are present in foods but may be added or consumed separately after chemical synthesis or production by microorganisms.

Food quality refers to the composite characteristics which differentiate individual, units and enable determination of the degree of acceptability by the consumer and It’s defined as the measure of the degree of excellence. Consumption of unsafe, contaminated food leads to food borne diseases which cause morbidity and mortality. Microbiological quality of food comprises of these aspects: food safety, shelf-life and consistency. A food should not contain pathogen or its toxin and the total microbial load associated, should be within limits, so that it is not organoleptically spoilt in a short period of time. A food must be of consistent quality both with respect to safety and shelf-life. The principal knowledge of food microbiology-helps in assuring a supply of wholesome and safe food to the consumer.

The applications help in knowing the effect of processing and producing economically and consistently foods; which have good keeping qualities and are safe to eat. However, the control of processing and environment through Good Manufacturing Practices (GMP) and Hazard Analysis Critical Control Point (HACCP) is of greater importance than even the examination of the end-product. Other than traditional methods to detect food-borne pathogens which use culture media to select and propagate viable cells in food, there are rapid methods which comprise of biochemical tests, specialized medial substrate and DNA and anti-body based detection assays. However as they are used mainly for preliminary screening, they should be combined with the traditional routine analysis.

Shelf Life

Foods can be placed in three groups according to the shelf life:

- Stable or nonperishable foods: These foods do not spoil unless handled carelessly and include products such as sugar, flour and dry beans.
- Semi-perishable foods: If these foods are properly handled and stored, they remain unspoilt for a fairly long period e.g. potatoes, garlic, some varieties of apples etc.
- Perishable foods: This group includes most important daily foods that spoil readily unless special preservatives methods are used. Meats, fish, bread, bun, cakes, poultry, most of the fruits and vegetables, eggs and milk belong to this category.

The shelf life of any food depends on the physical state, structure of the food, chemical properties of the food and the environmental conditions of its production, handling, processing if any and storage.

Causes of Food Deterioration

Food can be deteriorated by the following factors:

- Growth of microorganisms.
- Insects.
- Pests.
- Rodents.
- Birds.
- Action of enzymes of the plant or animal food.

- Purely chemical reactions i.e. those not catalyzed by enzymes of the tissues or of microorganisms.
- By physical changes such as those caused by freezing, burning, drying, pressure etc.

Chemical Reaction

The chemical composition of a food determines how satisfactory it will be as a culture medium for microorganisms. Each organism has its own characteristic ability to utilize certain substances as a source of energy, a carbon source and a source of nitrogen.

Chemical Changes in stored food take place at rates governed by its temperature and moisture content. The effect of temperature is such that a 10 °C rise, within the temperature range likely to be encountered in temperate or tropical conditions, causes an approximately two fold increase in reaction rates. Cool storage retards changes such as fat oxidation, vitamin loss and container corrosion. Foods such as dried milk benefit from even a small reduction in storage temperature. Keeping such foods cool and dry will usually greatly reduce the rate of development of brown discoloration ('browning') and the associated off-flavours.

The chief factors which influence microbial activity in food are: hydrogen-ion concentration, moisture, oxidation-reduction potential, nutrients and the presence of inhibitory substances every microorganism has a minimal, maximal and an optimal pH for growth. Yeasts and molds require low pH (below 4.5) for growth while bacteria are favoured by neutral pH. Microorganisms have an absolute demand for water without which no growth can occur. The water requirement is best expressed in terms of available water or water activity (aw). It is the vapour pressure of the solution (of solutes in water in most foods) divided by the vapour pressure of the solvent usually water. For pure water aw is 1.0.

Table: The values permitting growth of spoilage organisms.

Group of Microorganisms	Minimal aw value
Bacteria	0.91
Yeasts	0.88
Molds	0.80
Halophillic bacteria	0.75
Xerophillic (dry conditions) fungi	0.65
Osnophillic yeast	0.60

In general bacteria require more moisture to grow than yeasts and molds.

Biochemical Reaction

1. Putrefaction: Most of the nitrogen in foods is in the form of proteins which is hydrolyzed by enzyme protease increased by proteolytic microorganisms to a number of compounds, polypeptides, simpler peptides or amino acids before they can serve as nitrogenous food for most organisms. Anaerobic decomposition of proteins, peptides or amino acids result in foul smelling, sulfur containing products such as hydrogen, ammonia, amines, indole etc.

Protein + Proteolytic microorganisms = Amino acids + amines +NH_3 + H_2S.

2. Carbohydrate fermentation: Carbohydrates are preferred by microorganisms to other energy-yielding foods. They are fermented by micro-organisms to form acids, alcohol and gases. Yeasts cause alcoholic fermentation under anaerobic conditions with ethanol and carbon dioxide as the principal products. Lactic acid bacteria bring about lactose (milk sugar) fermentation yielding lactic acid as the main product. Coliform bacteria carry out coliform type of fermentation with chief products as lactic acid, formic acid, ethanol, carbon dioxide, hydrogen acetoin, butanediol etc.

Carbohydrates + action by microorganisms = Acid + Alcohol + Gases.

3. Rancidity: Fats are hydrolyzed by microbial lipase produced by lipolytic microorganisms to produce glycerol and fatty acids. Sometimes auto-oxidation of fats is more common. This is called rancidity.

Fat + Lipase = Fatty acids + glycerol.

Microorganisms: General Principles, Causes and Growth

Microorganisms important in food are moulds, yeasts, bacteria, ahtinomycetes and viruses. They are microscopic in size but arc easily visible to the naked eye when large numbers for colonies or surface films, except for viruses. Microorganisms ID foods can be either beneficial as ID fermented foods such as yoghurt, or harmful causing spoilage and disease. Deterioration of packaging and sealing materials due to mould growth often results in spillage; damp and mouldy cereal grain can become caked and difficult to handle; even the structure of stores can be weakened by certain microorganisms. The significance of microorganisms in stored food commodities depends upon a number of conditions:

- The types of microorganism.
- The number present.
- The type of food commodity and the treatments to which it has been exposed.
- Whether the foodstuff will be cooked.
- Who will consume the food.

General Biology

Recognition of all the microbial species that cause spoilage is not possible without a considerable amount of experience and laboratory equipment. However, adequate field recognition of the likely causes of spoilage can often be achieved by simple observation.

- Bacteria are the smallest of the microorganisms normally associated with food spoilage. They may be seen as slimy colonies on some spoilt commodities.
- Actinomycetes are similar to bacteria. Actinomycetes will often impart a white, speckled appearance to the grain. Thermoactinomyces spp. grows on grain which heats up in store. Some species have been associated with the production of carcinogenic compounds in dried fish products.

- Molds and yeasts are part of larger group of microorganisms called fungi. Molds frequently produce a cottony or velvety appearance on spoilt commodities and are also often associated with discolouration of cereal products.

Table: Types of food spoilage caused by micro-organisms.

Alteration of colour or texture	
Discolouration	Fungal/bacterial pigments, chemical changes, browning reactions due to microbial heating
Caking	Fungal growth
Physical deterioration	Growth of fungi, bacteria and actinomycetes
Unpalatability	
Fermentation	Acid and gas produced from carbohydrates
Putridness	Objectionable flavours and odours produced from protein breakdown.
Rancidity	Objectionable odours and flavours produced by fat hydrolysis and oxidation.
Production of toxic chemicals	
Carcinogens	e.g., aflatoxins, formed by fungi in the food
Toxic chemicals	Produced by micro-organisms that may cause chronic or acute toxicity if eaten.
Transmission of disease	Due to the presence of pathogenic bacteria

Molds

They are aerobic, i.e. they require-oxygen for growth and most of them require acidic pH for growth. Most of the molds possess a variety of hydrolytic enzymes. Molds of industrial importance are:

- Mucor sp. (produces amylase and used in making oriental foods).
- Rhizopus sp. (bread mold): It causes spoilage of many foods, berries, fruits, vegetables, bread etc.
- Thamnidium: It is found on meat in chilling storage.
- Aspergillus: The aspergilli are very widespread. Many are involved in the spoilage of foods and some are useful in the preparation of certain foods. Aspergillus 'tiger is used for commercial production of citric and gluconic acids and enzymes.
- Pencillium: This genus is also widespread. P. expansum (blue-green spored mold) causes soft rot of fruits. P. camembertii and P. roquefortii are used in ripening of camembert and roquefort cheese.
- Alternaria: They cause spoilage of many foods. Alternaria citri causes rotting of citrus fruits.
- Helminthosporium: Many are plant pathogens but may grow saprophytically on vegetable materials.

Yeasts

Yeasts may be useful or harmful in foods. Yeast fermentations are involved in the manufacture of foods such as bread, beer, wines, vinegar and surface-ripened cheese. Yeasts are also grown for enzymes and for food. Yeasts are undesirable when they cause spoilage of sauerkraut, fruit juices, molasses, hogey, meats, wine, beer and other foods. Yeast cells are spherical to ovoid in shape and reproduce asexually by budding or fission. Yeasts of industrial importance are:

- Saccharomyces: Saccharomyces cerevisiae used in leavening of bread, as top yeast for ale, production of alcohol, glycerol and invertase. S.cerevisiae var. ellipsodeus is used for production of industrial alcohol, wines and distilled liquors.
- Zygosaccharomyces: They are osmophilic and are involved in the spoilage of honey, molasses etc.
- Pichia and Hansenula sp: They form a pellicle on beers or wines.
- Debaryomyces: They form pellicles on meat brines and grows on cheese and sausages.

Fungal contamination of maize causing discoloration and caking of grains.

Bacteria

Two genera, Bacillus (aerobic) and Clostridium (anaerobic) fob endospores which may remain dormant for years and germinate under favourable conditions. Spores are formed due to nutrient depletion and are heat resistant structures (due to Dipicolinic acid) to tide over the unfavourable conditions. Some genera important in food are:

- Acetobacter: They oxidize ethyl alcohol to acetic acid. They are found on fruits, vegetables, souring fruits and cause spoilage of alcoho ic beverages.
- Aeromonas: It is a pathogen in humans, fish, frogs etc.
- Alcaligenes: It causes alkaline reaction and causes ropiness in milk.
- Bacillus: They are aerobic spore formers and may be mesophilic growing at moderate temperature or thermophilic (growing at above 45 °C temperature) Bacillus sp. is proteolytic and lipolytic and causes spoilage of many foods. Source is soil.

- Clostridium: They are anaerobic spore formers and many species actively ferment carbohydrates with the production of acids and gases (CO_2 +H_2). Source is soil.
- Escherichia: It is found in faeces and is pathogenic to humans.
- Halobacterium: They are strict halophiles (salt loving microorganisms) and cause discoloration on salted fish etc.
- Klebsiella: Klebsiella pneumoniae causes pneumonia in humans.
- Lactobacillus: They are microaerophilic (grow in very less O_2 concentration) and are rot shaped with ability to ferment sugars to lactic acid. They are used in production of fermented dairy products. They may also cause spoilage of wine, beer, refrigerated meats etc.
- Listeria: Listeria monocytogenes causes listeriosis.
- Mycobacterium: Mycobacterium tuberculosis causes tuberculosis and spreads mainly from raw milk from infected cows.
- Salmonella: They are enteric pathogens and cause food infections.
- Serratia: Serratia marcescens produces red discoloration in milk and on surface of foods.
- Shigella: They are transported by foods and cause bacillary dysentery.
- Staphylococus: They cause food Poisoning.
- Streptococus: Streptococus pyogenes causes human septic sore throat and scarlet fever.
- Vibrio: Vibrio cholerae causes cholera.
- Yersinia: Yersinia peslis causes human plague.

Viruses

Viruses are the smallest and simplest form of organisms. Unlike bacteria, viruses are not complete cells. Viral pathogens are often transmitted by infected food handlers through contact with sewage. Viruses do not require potentially hazardous foods to survive. Also viruses do not increase in number while they are in food, e.g. hepatitis A, Norwalk and Rota virus. Viruses contaminate food through poor hygiene by fold handlers, through a contaminated water supply or from sewage contaminated water.

Insects, Pests and Rodents

Insect and mite infestations cause weight loss and deterioration in quality by their presence and feeding activities. When insects are there in food, especially the grains, they produce heat, moisture, waste products and secretions. The heat, moisture and contamination have an adverse effect on the quality of grains and makes them susceptible to further deterioration. Increase in moisture content encourages the growth of molds. Infestation of the beetle Tribolium spp., produce large quantities of the chemical, benzoquinone, which is toxic and imparts an unpleasant taste and smell. The larvae of moths lips Ephetia sp. Plodia and Corcyra sp., bind flour and grain together into a solid mass. Thus insects and mites can lead to direct loss of food, decrease ill quality and results ill economic losses. Great care should be taken to prevent high value, processed foods e.g. biscuits and meals from being infested. It is impractical to remove insects from such foods so high standard of hygiene should be ensured in their preparation and storage.

Insects that attack grain are divided into two groups: primary pests and secondary pests, Primary pests are those that are able to attack and breed in previously undamaged cereal and pulse grains. Damage takes the form of holes and cavities within grains. Examples are Sitophilus sp., Rhyzopertha dominica, Sitotroga cercalella etc. Many of these species attack the commodity in the field prior to harvest. Secondary pests are not capable of successfully attacking undamaged grains and solid commodities. They are able to attack materials already infested by primary pests or damaged either by poor threshing, drying and handling. They are also able to attack processed commodities. Examples are Tribolium spp; Oryzaephilus spp, Corcyra spp; Plodia spp; etc. There are few species like Trogoderma Granariuni which do not fit in either category. Dried fish can be attacked by all pest species because its' surface has crevices. Primary pests are usually more destructive than secondary pests in short term storage of cereal grains. However on some commodities for example flour, milled rice, processed and blended foods, secondary pests, may form the majority of insects present.

Rodents such as rats and mice can attack stored foods. Apart from the food eaten, spoilt or contaminated, there are additional 'invisible' losses such as the replacement or repair of packaging materials and the cost of re-bagging spilled food. Rats and mice also feed inedible materials including electrical wiring. Hence their presence in buildings can constitute a fire hazard. Three main rodents are Rattusnorvegicus (Norway rat) Rattus rattus (Roof/ship rat) and Mus musculus (House mouse) and important diseases of man transmitted by them are Plague (Yersinia pestis), Rat bite fever (Spirillum minus), Food poisoning (Salmonella spp.), Leptospirosis (Leptospira spp.), Rickettsial Pox and Murine typhus (Rickettsia spp.) etc. Rodent control is essential for public health and can be done by the use of chemicals in food stores, trapping, proofing and improved hygiene.

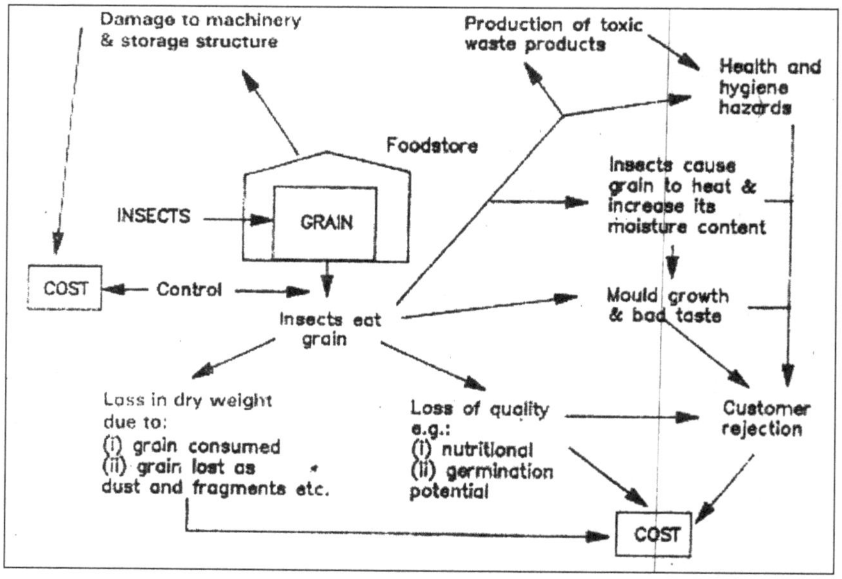

Economic loss caused by insect and mite infestations.

Nutritional Changes in Food

Food has a Variety of organic compounds in it. Numerous kinds of microorganisms can decompose them by hydrolysis caused by microbial enzymes, which lead to a decrease in the nutritional value of the food. There are three types of degradation processes:

- Protein biodegradation: Most of the nitrogen in foods is in the form of proteins which is hydrolyzed by enzyme protease elaborated by proteolytic microorganisms to a number of compounds, polypeptides, simpler peptides or amino acids before they can serve as nitrogenous food for most organisms.
- Carbohydrate biodegradation: Carbohydrates are preferred by microorganisms to other energy-yielding foods. They are fermented by microorganisms to form acids, alcohol and gases.
- Fat biodegradation: Fats are hydrolyzed by microbial lipase produced by lipolytic microorganisms to produce glycerol and fatty acids. Sometimes auto-oxidation of fats is more common.

Thus microorganisms when proliferate on food sometimes elaborate hydrolytic enzymes which leads to hydrolysis of organic constituents, thus decreasing their content. Hence we can say that these hydrolytic microorganisms decrease the nutritional value of food.

Food Borne Diseases

Organisms important in food include bacteria, yeast, molds and viruses. Food borne diseases result from the following:

- Bacteria
- Mycotoxins
- Viruses
- Rickettsias
- Parasitic worms
- Protozoa

Food Borne Intoxication by Bacteria

This refers food borne illness caused by presence of a bacterial toxin formed in the food. But there has to be large number of organisms (approx. lo^6/g of food) to produce toxin. They are of two types:

- Clostridium botulinum: This food poisoning (botulism) occurs rare but IS fatal. Symptoms due to the neurotoxin produced (7 types) is double vision, coated tongue, diarrhea paralysis and death within 3-6 days. The foods involved mainly are inadequately processed home-canned food or preserved meat and fish.
- Staphylococcus: The toxin produced is an enterotoxin which causes gastro enteritis or inflammation of the lining of the intestinal tract. The foods involved are cream filled bakery

goods, milk and milk products, meat and meat products, salads, puddings etc. ordinary cooking cannot destroy the toxin.

Food Infections by Bacteria

They are mild food poisoning and are of two types. First in those in which food is just a carrier of pathogens causing tuberculosis, diphtheria, typhoid, cholera etc. and second category in which food series as a culture medium for growth of pathogens like Salmonella, Vibrio, Escherichia coli, Clostridium perfringens, Yersinia Shigella listeria, Bacillus cereus etc. The symptoms are generally same, nausea, chills, diarrhea, cramps etc. and the number of organisms that can cause the disease are very small. The toxin causes gastroenteritus. Foods involved are fried nee, custards, cereal products, vegetable sprouts and mashed potatoes.

Mycotoxins

Mycotoxins are toxins produced by filamentous molds on foodstuffs. They are secondary metabolites that can cause death in humans and other animals. There are 300-400 compounds recognized as mycotoxins, out of which most common are aflatoxins and others are fumonisins, ochratoxin A, patulin, tricothecenes and zearalenone. Aflatoxins are produced by Aspergillus flavus and Aspergillus parasiticus and were first isolated from mold-contaminated peanuts in 1962, during which approximately 100,000 turkey birds died and the word mycotoxin was coined. Mycotoxins have been detected in cereals, oil-seeds fruits and vegetables, milk, bees, cocoa, raisins etc. both in field and in storage. Other molds which produce mycotoxins are Penicillium citrinum which produces citrinin, Fusarium produces Funonisins and zearalenone and Ochratoxin A from Aspergillus ochraceus. Mycotoxins can be used as chemical warfare agents. They can cause carcinogenicity, dermal irritation and other metabolic disturbances.

Aspergillus ochraceus causing discoloration and disintegration of wheat. Ochratoxin was isolated from the infected grains.

Ways of Controlling Mycotoxins

- Prevention of mold growth by proper drying and storage of food.
- Removal of mold damaged food material before storage or processing.
- Proper sanitation with adequate moisture and humidity control of food. A water activity of 0.85 and a temperature range of 25-40 °C should be avoided as they are optimum conditions for production of mycotoxins. Short periods of high temperature also should be avoided because it causes depression of growth of fungus and accumulation of toxin.

- Use of recommended anti-fungal substances when required.
- Regular chemical and bioiogical assays on food to detect mycotoxins.

Viruses

Viruses are ultramicroscopic parasites able to infect plants, animals and bacteria. There are two main viruses which are food borne: Poliovirus type causing poliomyelites, infections hepatitis virus causing jaundice and foods usually involved are milk, beverages, sea-food etc. Prevention can be done by maintaining personal cleanliness, adequate heating of foods and disinfection of water. The most important are human calciviruses (Norwalk like viruses) and the hepatitis A virus. Almost all viruses enter food as a result of faecal contamination. They cannot multiply in food. They can be inactivated by high temperature but can persist in refrigerated and frozen foods.

- Rickettsia: They are obligate parasites and like viruses they can be cultivated only on living tells. Rickettsia of public health concern is Coxiella burnetii which causes Q fever and can be excreted in milk in large quantities and can result in human infection. The pasteurization temperature of 62.8 °C (30 min.) in milk ensures its complete destruction.
- Parasitic Worms: They are occasionally encountered in food and are mainly flat worms, tape-worm or round worms. Flat worms, Fasciola hepatica are common in cattle and sheep, rare in humans and are caused by eating bovine liver. Tape worm (Taenia Solium can be present as contaminant in beef or pork. Round worms may be found in swine wild-game animals, under cooked fish etc. Thus cook fish, pork and beef thoroughly and avoid eating raw and under processed meat.
- Protozoa: The parasitic infections transmitted by food are caused by Entamoeba histolytica which causes amebic dysentery and the source is water contaminated with sewage or faeces. Gardiasis is caused by Giardia lamblia which occurs in vegetables washed with contaminated water and can occur in tap water or ice cream also. Toxoplasma gondil causes Toxoplasmosis and can be avoided by prevention of contamination of food and water from faeces.
- Toxigenic algae: A number of algae (dinofiagellates, diatoms and cyanobacteria) can produce very toxic compounds which may be transported to shell fish and a number of distinct illnesses are known such as paralytic, neurotoxic and diarrheal shellfish poisoning. These toxins are generally unaffected by cooling.
- Toxigenic fungi: These are some wild poisonous mushrooms like Amanita phalloides, Coprinus atramentarius, Psilocybe cubensis etc. which may cause gastro intestinal irritation, liver and kidney damage and sometimes even death.

Guidelines to Keep Food Safe

- Cooking/canning/heating should be adequate and the quality of raw ingredients added should be ensured.

- There should not be 3 lapses of 12 hours or more between preparation or eating.
- Keep meat fish, poultry and dairy products refrigerated or frozen. Refrigerate leftover promptly.
- Do not thaw frozen food at room temperature, use a microwave.
- Never taste or eat raw meat or poultry.
- Buy only clean, un-cracked eggs and use them within 5 weeks of purchase. Do not eat raw or uncooked eggs.
- Wash fruits and vegetables thoroughly even if you plan to peel them before eating them.
- Leftover food should be used within 2-3 days. If in doubt throw it out.
- Contaminated dish cloth can house millions of bacteria after a few hours. One option is to use paper towel to clean and throw immediately.
- Clean food and everything it touches thoroughly which includes washing food, utensils, counters, sinks etc.

Food Allergies and Poisoning by Chemicals

Food allergies may be caused by the following compounds:

- Histamines,
- Mono Sodium Glutamate (MSG),
- Nicotinic acid.

Poisoning by consumption of chemicals is rather uncommon and symptoms appear within a short time. Antimony, arsenic, cadmium, chlorinated hydrocarbons, copper, cyanide, fluoride, nicotinic acid, lead and zinc in foods can come from utensils, insecticides or fruit sprays.

Ergotism is caused by a fungus, claviceps purpurea. This fungus is a parasite of some grasses and cereals and produces a tough, purple brown sclerotium called ergot. Ergots contain alkaloid metabolites which are toxic and cause constriction of blood capillaries.

Anti-Microbial Agents used in Food

The anti-microbial agents used in food are grouped into the following three categories:

- Natural agents: They are not defined as such by law and include natural organic acid like lactic, citric and malic acids. They also include salts such as vinegar (acetic acid), sodium chloride, sugars, spices and their oils wood smoke, carbon dioxide and nitrogen.
- Agents generally regarded as safe (GRAS) for addition foods. They are listed in tables.
- Agents proved safe and approved by the Food and Drug administration.

The anti-microbial agents are added to food for inhibition or pilling of microorganisms and may be also classified according to their chemical composition, mode of action, specificity, effectiveness and legality. Sugar is effective because of its physical action while sodium benzoate due to its chemical reaction. While sodium chloride has a combination effect some anti-microbial agents are antiseptic in action rather than germicidal. Some are employed to treat wrappers or containers.

Table: Some GRAS anti-microbial agents.

Agent	Maximum Concentration allowed	Organisms affected	Foods
Benzoic acid/benzoates	0.1%	Yeast and molds. It is most effective at 2.5-4.0 pH.	Pickles, apple cider, soft drinks etc. Sodium benzoate is used extensively in jams, jellies, carbonated beverages, fruit juices etc.
Sorbic acid/sorbares	0.2%	Molds. It is most effective at low pH.	Used directly as an additive, as a spray, dip or coating on packaging material. Widely used in cheese, cheese products, baked goods beverages fruit, pickles etc.
Propionic acid/propionates	0.32%	Molds. They affect the cell membrane permeability.	Bread, cakes etc.
SO_2/Sulfites	200-300 ppm	Insects, microorganism. It cause reductions of disulfide linkages and inhibition of respiratory mechanism.	Wine making to reduce the normal flora of grape must, Molasses, dried fruits etc.
Ethylene/prophylene oxides	700 ppm	Yeast, Molds Ethylene oxide kills all microorganisms. They are strong alkylating agents attacking labile hydrogens.	Fumigant for spices nuts etc. Sterilants for packaging material and for cold sterilization of plastics, chemicals, syringes etc.
Sodium diacetate	0.32%	Molds.	Bread, cheese spreads, treatment for wrappers used on butter.
Nisin	1%	Lactics, Clostridia.	Certain pasteurized cheese spreads.
Sodium nitrite/nitrates	120 ppm	Clostridium botulinum.	Meat curing preparations.

- Acetic acid: Acetic acid in the form of vinegar is used in pickles, mayonnaise. It is more effective against yeast and bacteria than against molds and its effectiveness increases with a decrease in pH, which would favour the presence of the dissociated acid.
- Sugar and salt: Both lowers the water activity and thus have an adverse effect on microorganisms, Sodium chloride is used in brines and curing solutions or is applied directly o the food. It causes plasmolysis of cells and reduces solubility of oxygen in the moisture. Sugars make water unavailable to organisms.

- Alcohol: Ethanol, a coagulant and denaturize of cell proteins is most between 70-95% concentrations. The alcohol content in beer and wine etc. although cannot prevent spoilage but it, out limits the growth. Propylene glycol has been used as a mold inhibitor and as a spray to airborne microorganisms.
- Wood smoke: Wood smoke adds desired flavours and number of volatile compounds that have a bacteriostatic (arrests growth of bacteria) or bactericidal effect (pills bacteria). Fonnaldehuge is most effective of these compounds.
- Spices and condiments: They have an inhibitory effect on microorganisms. Cinnamon and cloves contain cinnamic aldehyde and eugenol respectively and are bacteriostatic. But the spices have to be treated to reduce their own microbial content. Onion and garlic have Acroleing and are bacteriostatic or germicidal.
- Antibiotics: Many antibiotics have been tested on raw foods, chiefly meat, fish and poultry in order to lengthen the storage time at chilling temperature. Nisin, subtilin, bacitracin etc. has been tried. The FDA has approved chlortetracycline (Sppm) for preserving poultry.

Enzyme Inactivation

Enzyme inactivation prevents product deterioration during storage. Most of the food and microbial enzymes are destroyed at 79.4 °C with some exceptions. Some - proteases and lipases retain their activity even after ultra-high-temperature process and can spoil the processed product during long-term storage. Detection of enzyme bovine phosphatase m processed milk indicates that the milk was not properly pasteurized, so it acts as a monitor to check the quality of milk. Blanching is a heat process applied to foods especially fruits and vegetables in which after washing, fruits and vegetables are heated at <100 °C (86-98 °C) for 1-5 minutes to inactivate, enzymes, for expulsion of oxygen from tissues and reduction in microbial load.

Treatments

Physical Treatment

Physical treatments are used to inhibit, destroy or remove undesirable microorganisms from food without involving chemicals. They include:

- Physical dehydration processes like drying and freeze-drying.
- Cooling or cold storage.
- Freezing and frozen storage.
- Heating (thermal treatment).
- Microwave heat treatment.
- UV or ionizing radiations.
- Non-thermal methods like high hydrostatic pressure pulsed electric fields etc.

1. Physical dehydration processes: Food preservation by dehydration is based on the principle that microbial growth is inhibited if the available water required for the growth is removed, thus reducing the water activity (aw). It can be done by-hot air, freeze-drying or freeze concentration m case of liquid foods, where freezing is followed by mechanical removal of frozen water.

2. Drying: During hot air drying of foods, the air stream has two functions, it transfers heat to evaporate the water from the raw materials, and it carries away the vapour produced. In the course of drying, microorganisms on the raw material are affected by both the drying temperature and changes of aw. Wet heat is more lethal than dry heat. The aw of dried foods is usually considerably lower than the critical value for microbial growth and hence are stable microbiologically, provided the relative, humidity of storage atmosphere less than 70%. The quality of raw materials used and their right handling should also be ensured.

3. Freeze Drying: Freeze drying (lyophilization) is gentle dehydration of the food in the frozen state through vacuum sublimation of its ice content, thus decreasing the aw.

4. Cool storage: Cool storage generally refers to storage at temperatures from 16 °C to minus f °C at which the rates of chemical reactions and growth of microorganisms are decreased. However psychotropic microorganisms can grow at cold teıperature such as Yersinia enterocoitica, Vibrio parahaemolytics, Listeria monocytogenes etc. and are important especially in minimally processed, extended shelf-life products. Variation of storage temperature should be avoided.

5. Controlled-Atmosphere Storage: (CAP) Controlled-atmosphere storage of certain fruits and vegetables is widely used in which there is reduced oxygen content (2-5%) an increased (8-10%) CO2 content in air tight chilled storage rooms. This modified storage atmosphere depresses the respiration, maintains the sensory and texture quality of perishables and also inhibits the growth of certain spoilage Organisms However in modified atmosphere packaging (MAP), which has a similar principle to CAP, one should ensure the absence of a pathogen, Clostridium botulinum which grows in the absence of oxygen.

6. Freezing and Frozen Storage: Freezing normally lowers the temperature of a -food to – 8 °C and the food is stored at this temperature or below. Many convenience food products are made by this technology. However, it is energy intensive. There are three basic freezing methods: (i) Cold air, (ii) indirect contact freezing (iii) submerging food or spraying the cold refrigerant liquid onto the food. At temperatures below about -8 °C there is practically no microbial growth. However the thawed frozen products are vulnerable to microbial spoilage and' refreezing of thawed products should be avoided.

7. Preservation by heat treatment: The most effective and most widely used method for destroying microorganisms and inactivating enzymes is heat treatment. The heat resistance of various microorganisms is very important from an industrial point of view. Pasteurization is a mild heat treatment aimed at inactivating enzymes and destroying 99-99.9 % of vegetative bacterial cells and to eliminate non-spore forming pathogenic bacteria, thus extending the shelf-life of food. Sterilization refers to complete destruction of microorganisms. Shelf-stable products may contain low number of viable but dorment bacterial spores and these products are called "Commercially Sterile". Commercial sterilization of food in hermetically closed cans or bottles is called canning or appertization. Exposure to ultra high temperature (140 °C), holding for several seconds and then

rapid cooling leads to shelf-stable foods at ambient temperature. The use of higher temperature for short time process leads to Iess nutrient damage. Thermal death time is defined as the time necessary to kill a given number of organisms at a specified temperature. The slope of the thermal death time curve is expressed by the Z value, which is the temperature increase required to reduce the thermal death time by a factor of 10. For the determination of heat processes for canned foods, F value is defined as the number of minutes at a specific number of viable cells with specific Z value. Thermal death point is the temperature necessary to kill all the organisms in 10 min.

8. Microwave Heating and Processing: This is becoming popular at the consumer level. Microwaves are electromagnetic waves between infrared and radio waves at 500 MHz to 10 GHz. The heat produced by the excitation bf food molecules results in microbial destruction.

Preservation by Irradiation

- UV radiation: The microbiologically most destructive wavelength range of UV radiation is between 240-280 nm at which nucleic acids of ells are damaged. Inactivation is caused by the cross-linking of thymine dimer of the DNA, thus preventing repair and reproduction. Gram negative bacteria, are most easily killed by UV, while bacterial endospores, molds and viruses are m eh more resistant. The UV sources are used mainly for disinfection of air and in as tic filling of liquid food, in cheese ripening and for disinfection of water.
- High-Intensity pulsed light: Decontamination of food surfaces with high-intensity xenon arc lamps is also used.
- Ionizing radiation: Food irradiation is limited to gamma ray of ^{60}FO or X rays with energies upto 5 MeV-10 MeV. The approximate dose to kill microorganisms varies from 0.5 - 200 KGy and for Insects IS 0.01 - 1 K by. 'The death of microorganisms by ionizing radiation is the result of damage to DNA.

Non-thermal physical preservation methods:

- High hydrostatic pressure: High hydrostatic pressure of 300-1000 MPa can be applied at ambient temperature and microbial inactivation occurs due to cell leakage caused by changes in proteins and cell membrane.
- Pulse electric field (PEF): High voltage discharges can kill microorganisms. An external PFE charges cells in such a way that they behave as small dipoles and an electric potential develops over the cell membranes. For vegetative bacteria the critical electric field strength is 15 KV/cm. The application or PEF has been tried in liquid foods and in meat industry.

Hygiene and Sanitation

Food safety should begin on the farm itself. From planting t consumption there are many opportunities for bacteria, viruses and parasites to contaminate the produce. On the farm, soil, manure, water, animals, equipment and workers may spread harmful microorganisms. Produce may be harvested on a farm, processed in one plant, repacked in another, then stored, displayed or served. Each, of the steps is an opportunity for harmful microorganisms to enter the food supply. Clean soil, clean water, air and surfaces and hygiene among workers, can minimize the pathogen contamination. But there is no way to guarantee that everything we grow and consume is free of harmful microbial contamination. The risk can be reduced if preventive steps are taken. Animal manure and human faecal matter represents a significant source of human pathogens and thus should be closely managed to limit the potential for pathogen contamination. Risk assessment should be done, HACCP approach should be used to assess the risks, identify critical control points, appropriate preventive measures, and to record the plant hygiene and to determine corrective measures in case of non-conformity, Bacteriological monitoring and record keeping should be there. Thus keeping microbial loads at minimum levels is essential to provide safe food of high quality.

HACCP Plan for Food-Processing

HACCP concept provides a systematic structured approach to assure the safety of a food product. However, there is no blueprint or universal formula for putting together the specific details of a HACCP plan. For it to work well, it should be a participatory endeavor at all levels of an organization, both in formulating and managing the p!;m' The integrity in the entire food chain cannot be guaranteed as a result of farm level vulnerability Hence farm level HACCP is also very important. Steps to control hygiene in food-processing:

- Site and infrastructure of food-processing: The site should be on cleared ground. It should be away from sources of insects/rodents/smells/compost piles/wells/water source. There should be road access for bringing in raw materials. The surrounding area should be planted with grass, as short grass acts as a very efficient trap for airborne dust. The building should be painted and tiled. Ceilings, doors, working surface and walls should be smooth, non-toxic and easy to disinfect. At the bottom of the wall where it meets the floor, there should be a right angle so that the dust is not collected. The windows should be such so that they are easily cleaned.

- Washing & toilet facilities: They should be separated preferably in a separate building or two closed doors should be there between the toilet and the processing area to prevent insects/odours. The toilet facilities should be fly proof.

- Drains: The drains should be covered by a fine mesh and drainage should be proper, floors (slope) and should be self-drained.

- Electric points: They should be fixed at least 1.0-1.5 m high on the walls to keep them dry.

- Building and equipment: They must be kept clean at all times. Equipment should be made up of smooth, non-toxic material and should permit adequate cleaning. Workers must be trained to keep equipment clean through the day and to remove wastes from the building. All dust from high window ledges/old sacks etc. should be removed so as to avoid breeding of insects.

- Sanitization: Sanitization can be done by the use of bleaching powder i.e. (30-40 g/litre) 1%, 2 ml in 100 ml. (200 ppm), 1 ml in 100 ml for processing table, utensils and equipments for floors the concentration required is 200 ppm and for drains it is 250-500 ppm.

- Processing room: No food should be eaten in the processing room.
- Water Supply: The water supply should be potable/WHO guidelines and, cater to the standards specified for washing and drinking purpose. Non-potable water should not be allowed to mix into the potable water systems.
- Air quality, ventilation & lighting: All of them should be adequate as to avoid food contamination. Natural or mechanical ventilation should be provided to control air-borne contamination of food and control of o1ours and humidity. Adequate natural or artificial light should be provided to ensure that food is not contaminated by breakages.
- Storage: Adequate storage facilities should be provided which allow effective cleaning and suitable environment to minimize deterioration of food. Raw material & processing area should be separate as to avoid cross contamination. All extremely opening doors and should be fitted, with automatic air curtains. There should be fly proofing of all windows, ventilators, drain outlets/holes should he netted. The fly catchers should be away from food handling points.
- Personal hygiene: The infected workers should be excluded from work and all workers should undergo periodic training on the important role of personal hygiene in the quality of the processed product. Adequate means of hygienically washing and drying of hands and suitable toilets should be provided.

Training

Food handlers should be trained in the basic concepts and requirements of food and personal hygiene as well as those aspects particular to the specific food processing operation. The level of training varies depending on the type of operation and the precise job description of the employee however some form of induction training with regular updating or refresher courses IS an absolute minimum.

Training should give food handlers an understanding of the basic principles of hygiene, why it is necessary, and how to achieve it in practice. A core curriculum for any such course should emphasize:

- Micro-organisms as the main cause of food spoilage and food borne illness and the characteristics of the common types of food poisoning.
- How to prevent food poisoning through the control of microbial growth, survival or contamination.
- Standards of personal hygiene required of food handlers. These are principally to avoid contamination of food with bacteria. The food handler may harbor as part of the body's flora, e.g. Staph, aureus, Salmonella or that they may bring in with them from the outside world, e.g, Listeria, B. cereus.
- Principles of the handling and storage of foods such as the correct use of refrigerators and freezers, the importance of temperature m9nitoring, the need for stock rotation and the avoidance of cross-contamination.

- Correct cleaning procedures and the importance of the 'Clean-as-you-go' philosophy.
- Knowledge of the common pests found in food premises and methods for their exclusion and control.
- An introduction to the requirements of current food legislation.

Some Do's and Don'ts of Personal Hygiene for Food Handlers

Do's

Wash your hands regularly throughout the day and especially:

- After going to the toilet.
- On entering a food room and before handling food or equipment.
- After handling raw foods.
- After combing or touching the hair.
- After eating, smoking, coughing or blowing the nose.
- After handling waste food, refuse or chemicals.
- Keep fingernails short and clean.
- Cover cuts, pots or boils with a waterproof dressing.
- Keep hair clean and covered to prevent hair/dandruff entering the food.
- Always wear clean protective clothing (including footwear) in food processing areas.

Don'ts

- Do not smoke, chew gum, tobacco, betel nut, fingernails or anything else.
- Do not taste food.
- Do not spit, sneeze or cough over food.
- Do not pick, nose, ears or any other body site.
- Do not wear Jewelry when handling food.
- Do not wear protective clothing outside the production areas.

References

- Food-processing-and-preservation-journals-articles-ppts-list: scitechnol.com,
- Retort-processes: retorts.com, Retrieved 13, August 2020
- Processing-radio-frequency-processing-of-food, food-technology-magazine, news-and-publications: ift.org, Retrieved 04, March 2020
- Microwave-Application-for-food-Processing-264352691: researchgate.net, Retrieved 15, June 2020

Chapter 3

Food Packaging Technology

The technology involved in the packaging of food for the purpose of protecting it from tampering as well as fulfilling its physical, chemical or biological requirements falls under the umbrella of food packaging technology. The different aspects of food packaging technology such as aseptic packaging, active and intelligent packaging, and packaging of dairy products have been thoroughly discussed in this chapter.

In today's world, packaging is universal and important too. It surrounds, enhances and protects the goods we buy, from processing and manufacturing through handling and storage to the final consumer. Without packaging, materials handling would be a difficult, inefficient and costly exercise and modern consumer marketing would be virtually impossible. Most of the containers in the market today are used to protect a specific quantity of product during procurement, storage, distribution and retail sales, although several are also designed for bulk supply. The quality of the individual package depends on the nature, uniqueness and value of the product besides the prevailing social practices and legislation.

The selection of a packaging, storage and distribution system will depend on existing economic ability, production and distribution efficiency, retailing pattern, consumer preferences and ecological aspects. Despite the importance and key role which packaging plays, it is often regarded as a necessary evil or an unnecessary cost. Furthermore, in the view of many consumers packaging is, at best, somewhat unnecessary and at worst, a serious waste of resources and an environmental threat. Such an opinion arises because the functions which packaging has to do are either unknown or not considered fully. By the time most consumers come into contact with a package, its job in many cases is almost over, so it is understandable that the view that excessive packaging has been used has gained some belief.

The Packaging Institute International defines packaging as the enclosure of products, items or packages in a wrapped pouch, bag, box, cup, tray, can, tube, bottle or other container form to perform one or more of the following functions: containment; protection and/or preservation; communications; and utility or performance. If the device or container performs one or more of these functions it is considered a package.

Functions of Packaging

The functions of a package are to preserve the quality and freshness of food, to add appeal to the food to attract consumers, and to facilitate its storage and distribution. The basic functions required of a package can be grouped under five major categories.

To Contain the Product

The primary function of any package is to contain the food and facilitate handling, storage, and distribution all the way from the manufacturer to the ultimate user or even the time the rest portion is utilized by the consumer. However, there are usually various levels of packaging. A primary package is one that comes into direct contact with the contained product, e.g., metal cans, glass jars, and plastic pouches. By law, a primary package must not yield any substance that may be injurious to the health of the consumer. Further development to facilitate handling is to bundle a series of primary packages together, and this lead to the concept of secondary packages. Examples of secondary package is corrugated box in which tins of apple juice are packed. As methods of handling and transportation have become more sophisticated, these secondary packages are often palletized and secured by strapping with metal or, more commonly, by shrink- or stretch-wrapped film to give yet another level of packaging, i.e tertiary packaging. In turn, these pallet loads may be packed into large metal containers, i.e., quaternary packaging for transportation over long distances by air, land, or sea. The secondary, tertiary and quaternary packaging is also known as packing. The following are basic functions during containing.

- Adequate size and shape (trays to support biscuits in package).
- Proper constructional features. No leakage, spillage, diffusion, i.e. loss prevention.
- Package: Must contain the commodity in natural form (chips packed in Pillow pack, prevent damage).
- No subsequent damage after packaging during handling transportation and storage.
- Optimum compatibility (nontoxic, non-soluble with product, No physical, chemical or biochemical changes/alteration, i.e. inert to the product.)
- Containment or agglomeration: Small objects are typically grouped together in one package for reasons of efficiency. For example, a single box of 1000 pencils requires less physical handling than 1000 single pencils. Liquids, powders, and granules need containment.

To Protect the Product

One of the most important functions of any container is to protect the product contained against any form of loss, damage, deterioration, spoilage, or contamination that might be encountered throughout the distribution chain. Packaging can prevent physical damage, e.g., bruising caused by vibration shocks during transportation or stacking in a warehouse. Proper packaging will also prevent material loss, e.g., potatoes from a weak sack or juice from a leaky can. Packaging can also protect products against moisture loss or gain, dust, and light, which causes deterioration of some light-sensitive products. It can also protect the package contents against temperature fluctuations in the transit of chilled and frozen foods. Packaging can also be used to control the availability of oxygen to fruits and vegetables and to protect against loss of flavor or fragrance and help products retain their nutritional value. Proper packaging may also protect the product against microbial spoilage by bacteria, yeasts, and molds. It can also protect against microbiological spoilage of stored products due to rodents and insects.

Packaging protects the product against damages which may be due to different hazards viz. (a) Mechanical, (b) Environmental (c) Microbial and Biochemical and (d) Social.

Table: Hazard, damage and protection of packaging materials.

Sr. No	Storage	Hazard	Damage	Protection
I	Handling and transportation	Drop, shunting, shocks, vibrations, stack load, compression etc.	Breakage, loss of shape, dusting, seepage.	Cushioning, blocking.
II	Storage	Stack load, compression, Attack by rodents and insects.	Crushing, distortion sticking, spillage, contamination, spoilage.	Adequate compression strength of package. resistance and repulsiveness to insects.
III	Environment during storage transportation and distribution	Biological or otherwise.	Contamination.	Toughness of packaging material (to resist penetration).
		High/low humidity moisture/water.	Physical, chemical and biological deterioration due to loss/gain of moisture.	Efficiency of closure providing. Water vapour barrier properties. Package desiccant etc.
		O_2	Oxidative rancidity.	O_2 BARRIER VACUUM – O_2 N_2/CO_2 flushing.
		Light U.V. rays	Vitamin Destruction, Off flavour development, Oxidative rancidity, Bleaching of pigments.	Use of opaque or dark coloured packaging material.
		Temperature	Change of state, Increase of moisture ingress Increased rate of deterioration.	Heat insulation Use of poor conductor Use of reflective insulation.
		Time	Gradual and slow changes occur and staling and other deteriorative changes occur.	Early/immediate marketing (FIFO) Proper schedule of dispatching order providing Heat insulation Use of Barrier material.

Barrier protection: A barrier from oxygen, water vapor, dust, etc., is often required. Permeation is a critical factor in design. Some packages contain desiccants or Oxygen absorbers to help extend shelf life. Modified atmospheres or controlled atmospheres are also maintained in some food packages. Keeping the contents clean, fresh, and safe for the intended shelf life is a primary function of the package.

Medium of Information

An important function of any food package is to identify the product and its origin; to inform the consumer how to use the contents; to provide any other information needed or required; and very importantly, to attract the user and encourage purchase of the product. Package design has been an important and constantly evolving phenomenon for many years. Marketing communications and graphic design are applied to the surface of the package and in many cases the point of sale/display. The information a package can convey to the consumer may include the following:

- Product manufacturing and best before dates.
- Proper storage conditions.

- Instructions for use.
- Size and number of servings or portions per pack.
- Nutritional information per serving.
- Manufacturer's name and address.
- Cost.
- Suggested recipes.
- Country of origin.
- Information transmission: Packages and labels communicate how to use, transport recycle, or dispose of the package or product.

Means of Minimizing Costs

An important factor often overlooked is that packaging actually reduces costs for the consumer. Packaging reduces food costs by reducing the cost of processing. Foods can be processed where they are grown, waste is treated at the processing plant, and shipping weights are reduced, thereby lowering the cost of transportation. The handling of packages in quantity is important for the economics of bulk storage, warehousing, transport, and distribution. Proper packaging facilitates efficient and mechanized handling, distribution, and marketing of products, thus reducing the high labour costs that would have to be absorbed into the price of the product. Thus, packaging not merely contains the product, but it is a process of bringing goods from the production point to the point of use in a most beneficial manner. This involves all aspects of handling, storage, preservation, distribution, advertising, sales promotion, preparation and various other facts of industry.

Means of Selling Product

The packaging and labels can be used by marketers to encourage potential buyers to purchase the product. Packaging is often referred to as the silent salesman.|| Robertson concisely summarized the multifunction of packaging when he stated that a package must protect what it sells and sell what it protects.|| Packages can have features which add convenience in distribution, handling, display, sale, opening, reclosing, use, and reuse.

According to Jelen, primary packages should have the following characteristics to facilitate the sale of products:

- Aesthetic appeal.
- Non-toxic.
- Transparent.
- Lightweight.
- Tamper evident.

- Easy to pick up and handle.
- Easy to fit into cupboards, shelves, refrigerators, etc.
- Easy to open and dispense from.
- Easy to reclose.
- Returnable, recyclable, or reusable.
- Safe and presents no hazards in the way of broken glass or sharp jagged metal edges.
- Display the product.
- Glamorize: Create an illusion of something very precious, by decoration, embossing techniques and exotic closures, but it should not deceive the people.

The desirable polyfunctional properties of packaging materials are summarized in Table.

Table: Functional Requirements of Packaging Materials.

No.	Functional Property	Specific Factors
1	Gas permeability	O_2, CO_2, N_2, H_2O vapor.
2	Protection against environmental factors	Light, odor, microorganisms, moisture.
3	Mechanical properties	Weight, elasticity, heat-sealability, mechanical sealability, strength (tensile, tear, impact, bursting).
4	Reactivity with food	Grease, acid, water, color.
5	Marketing-related properties	Attractiveness, printability, cost.
6	Convenience	Disposability, repeated use, resealability, secondary use.
7	Aroma	Aroma compound barrier property.

Other Functions of a Package

- Dispensing: Product not used all at once, remove a portion, without destroying/damaging the remaining product/container.
- Preserve: Remaining product in container-protection and preserve it for extended/desired period.
- Measuring/Portion control: Single serving or single dosage package has a precise amount of contents to control usage. Bulk commodities (such as salt) can be divided into packages that are a more suitable size for individual households. It also aids the control of inventory: selling sealed one-liter-bottles of milk, rather than having people bring their own bottles to fill themselves.
- Security: Packaging can play an important role in reducing the security risks of transport. Packages can be made with improved tamper resistance to deter tampering and also can have tamper-evident features to help indicate tampering. Packages can be engineered to help reduce the risks of package pilferage: Some package constructions are more resistant to pilferage and some have pilfer indicating seals. Packages may include authentication

seals to help indicate that the package and contents are not counterfeit. Packages also can include anti-theft devices, such as dye-packs, RFID tags, or electronic article surveillance tags, that can be activated or detected by devices at exit points and require specialized tools to deactivate. Using packaging in this way is a means of loss prevention.

Packaging Types

Terms Used

- Package: It cuts contact between material and outside influences. Package material comes in direct contact with the product (Packaging).
- Pack: Secondary container. Packing material never comes in contact with product.
- Packing: Number of containers/packages put together in big container is called packing.

Packaging may be looked at as several different types. For example a transport package or distribution package is the package form used to ship, store, and handle the product or inner packages. Some identify a consumer package as one which is directed toward a consumer or household. It is sometimes convenient to categorize packages by layer or function: "primary", secondary", etc.

- Primary packaging is the material that first envelops the product and holds it. This usually is the smallest unit of distribution or use and is the package which is in direct contact with the contents (viz. butter in parchment paper).
- Secondary packaging is outside the primary packaging – perhaps used to group primary packages together (viz. paper board pack containing butter wrapped in veg. parchment paper).
- Tertiary packaging is used for bulk handling, warehouse storage and transport shipping. The most common form is a palletized unit load that packs tightly into containers (viz. Boxes containing 20-25 or 50 butter packs are put together).

These broad categories can be somewhat arbitrary. For example, depending on the use, a shrink wrap can be primary packaging when applied directly to the product, secondary packaging when combining smaller packages, and tertiary packaging on some distribution packs.

Table: Differences between packaging and packing.

No.	Packaging	Packing
1	Comes in direct contact with the product.	Never in direct contact.
2	Called primary packaging material.	Secondary/Tertiary/Quaternary.
3	Should be food grade, non-toxic, tasteless, odourless, lowest possible migration.	No strict requirements.
4	Packaging- a must e.g. Ice cream party pack, Bulk pack, Ghee.	May be done/may not be done. Packaging then packing e.g. CFB, cartons, etc. Bulk biscuit packs.
5	Materials used: Plastics/glass/metal/ treated paper or their combination.	CFB/Plastic board boxes, wood, metal, etc. Shrink/ stretch wrapping.

6	Objectives: Mainly to contain, carry, protect. Help in selling, legal aspects, marketing/sale, technical, transportation.	Mainly ease in transportation and protection of packages.
7	Generally attractive. Not a must: Biscuits and rolls in a pack. E.g Kellogs flakes, toffee. The exposed portion must be attractive.	Generally not attractive. But if retail pack, secondary packing exposed to consumers then attractive: Butter carton.
8	Recycled material never used.	Much preferred.
9	Selection of packaging material: Physicochemical properties of product are considered.	Generally stress/strength properties puncture resistance/burst strength, folding endurance, environmental factors considered.
10	Keeping quality is determined by packaging material.	Generally not so.
11	Single unit packaging.	Generally multi unit packaging. Sometimes single unit also. Butter carton, Bag in box. Here packing materials should be more attractive/effective than packaging material.

Reasons for selecting a particular style/type of packaging are vast and varied, numerous and changing. Product and packaging are becoming so interdependent that one cannot separate/consider one without another. Greatest part of food is spent in some form of package.

Requirements for Producing Successful Package

Some sets of facts are necessary to be known for producing a successful package (mainly related to product - package interaction and transportation):

Facts about the Product

- The nature of the product, the material from which it is made and the manner in which it can deteriorate.
- Its size and shape.
- Its weight and density: Eg. powder bulk density, size of tins.
- Its weakness: Which parts will break, move about, become bent or scratch or abrase the box easily.
- Its strengths: Which part will withstand loads or pressures and which might be suitable for loading the product in the pack.
- The effect of moisture and temperature changes on the product and whether it will absorb moisture or corrode.
- Compatibility: Whether the product is likely to be affected by any of the possible packaging materials, which items can be packed together, with protection if necessary and which items must not be packed together under any circumstances.
- How far stripping down may be carried out to reduce the package size to a minimum such that the customer can handle them.

Facts about the Transport Hazards

- The type of transport: Road, rail, sea or air.
- The degree of control over the transport. Is it private or public transport?
- The form of transport: Bulk, freight container, unitized load, postal, passenger train, etc.
- The mechanical conditions and duration of storage (Manufacturer → State Distributor → District Distributor → Taluka/City → Retailer. The longer the journey or handling more strength is required in packaging and packing materials leading to higher cost).
- The nature and intensity of mechanical and climatic hazards in transport, storage, retailing and use. Packaging/packing material has to withstand wide range of temperatures and relative humidity.
- Whether handling aids are available for loading and off-loading at all points between maker and user. (Viz. Lifts, Trolleys, Slip conveyers etc.)
- The importance of minimum volume in relation to transport costs. Over packaging must be prevented.

Hazards May be

- Mechanical: Impact (vertical, horizontal), stationary package impacted by another, vibration, compression, Racking or deformation, piercing, puncturing, tearing etc.
- Climatic hazard: (High/low temperature/pressure) light, liquid/water (fresh/polluted), dust, and water vapour, R.H.
- Biological: (Microorganisms, fungi, moulds, bacteria, beetles, moths, flies, ants, termites, mites, rodents (rats and mice), birds.

Contamination by other Goods

- By materials of adjacent packs.
- By leaking contents of adjacent packs.
- Radioactivity.

Packaging Equipment and Machinery

When we consider machines for packaging, there is an endless variety of packaging and processing machineries for food products, and many of them are remarkably similar. One of the major problem in selection of machinery is distinguishing which, if any, offer a specific advantage over the other. Also, the degree of versatility may have to be considered; it is pointless to purchase a do-everything machine when there is a single task at hand, similarly machine with too little adaptability may be a blockage in variable production process.

Intermittent and Continuous Machinery

Machinery can operate on an intermittent or continuous basis depending on the requirements of the production and the amount of capitalization. Normally, intermittent machinery will take a single package or small number of packages executes an operation on them all at once, and then pass them to the next process step. A continuous operation machine receives a stream of packages and executes the operation without stopping or reducing the speed of the overall flow of material. However, a four station manual-feed filling machine might be suitable for a small winery, that machine would be useless for a soft-drink bottler who needs several thousand fillings per minute to remain competitive. Equally, while continuous machinery is more economical at high throughput rates, it might make less sense to have a high-capacity operation that is in operation only for several hours a week because of low demand, although for the smallest of markets this may happen because of the employment of used machinery and the lack of properly scaled equipment.

Characteristics and Classification of Machinery

Choice of packaging machinery depends on:

- Technical capabilities, labour requirements.
- Worker safety.
- Maintainability.
- Serviceability.
- Reliability.
- Ability to integrate into the packaging line.
- Capital cost.
- Floorspace.
- Flexibility (change-over, materials, etc.)
- Energy requirement.
- Quality of outgoing packages.
- Qualifications (for food, pharmaceuticals, etc.)
- Throughput.
- Efficiency.
- Productivity.
- Ergonomics, etc.

Packaging machines may be classified in to following general types:

- Blister packs, skin packs and vacuum packaging machines.

- Bottle capping equipment, over-capping, lidding, closing, seaming and sealing machines.
- Cartoning machines.
- Box, case and tray forming, packing, unpacking, closing and sealing machines.
- Cleaning, sterilizing, cooling and drying machines.
- Conveyors, accumulating and related machines.
- Feeding, orienting, placing and related machines.
- Filling machines: Handling liquid and powdered products.
- Package filling and closing machines.
- Form, fill and seal machines.
- Inspecting, detecting and check weighing machines.
- Palletizing, depalletizing, unit load assembly.
- Product identification: Labeling, marking, etc.
- Wrapping machines.
- Converting machines.

Other specialty machinery include slitters, perforating machines etc.

Types of Packaging Machinery

Fillers, Feeders, Metering and Measuring Machines

The job of fillers and feeders is to take product in a determined number or amount to fill the assembled package. There is a wide spectrum of different types of equipment to for this purpose, many of which can handle difficult materials such as thick pastes and creams, or very fine objects.

The simplest of them will fill and dump baskets, cups, or trays of material, are well suited to loose and granular products such as cereal and animal food, and can be used for single objects if the objects are properly guided in feed chutes going to the packaging machine. These may be fitted with weighing sensors to permit a net-weight fill to be directly applied to the product, and may use a combination of small fill trays to estimate the net weight more accurately. It is possible to use auger metering where the product is dispensed by based on timing or counting the rotations of an auger that delivers product into the package.

Many types of powders, particulates, and granular products are subjected to bridging and arching phenomena, where the powder will jam feed chutes if they are not suitably designed for such product. Fine powders practise cohesive arches using inter-particulate bonding whereas larger particles will make interlocking arches where the mechanical locking of irregularly shaped particles provides the cohesiveness to form arches. Both of these phenomena depends on the same principles that traditional arch structures in bridges and buildings use – transferring circular compression into vertical and horizontal force to maintain the arch structure. Proper design of the hoppers for the product as well as adaptive equipment such as agitators or bridge breakers may be necessary to keep the device operating properly.

On the contrary bridging phenomena can be used to stop product from flowing in certain types of measuring devices, such as screw-auger fillers. They are designed with a small disk at the end of the auger that creates blockages when the auger is not rotating and depends on bridging to stop the flow between package fillings.

Direct counting machines can use trays, simple photocells, or complex vision-system pick and-place manipulators to acquire the exact number of products required in a package. Similarly, it is possible to use the package or an inner tray as a measuring device, filling it with the appropriate number of product units, and then enclosing it in a larger secondary package.

Filling by Gravitation

Filling by gravity is used for filling thin liquids like milk into glass bottles/plastic bottles. In gravitational filling, the filling process is stopped when the pre-calibrated filling height has been reached. This system is suitable for filling milk in glass bottles. However, packaging milk in bottles is outdated. In volumetric filling process, fixed volume of liquid is filled.

Mechanical Filling

In this type milk powders are metered filled by using screw conveyors. The conveyors are used as metering and dosing devices. However, due to variation in bulk densities of milk powder, care must be exercised to ensure even delivery of the product and which shall be equal in weight from package to package.

Cup Filling Machines

Plastic cups either preformed, from a film in the machine itself or readymade cups when used are placed in the stacker of the machine. Automatic packaging machines which work aseptically are in use for the production of long life products. In these machines the film is passed through a bath of Hydrogen peroxide and then goes to a sterile tunnel in which it is sterilized by either exposure to high temperature or exposure to excess pressure of air. The remaining stage like cup moulding, filling and sealing with lid takes place in the sterile tunnel.

Form-Fill-Seal Systems

Form-Fill-Seal (FFS) systems are preferred methods for many food items, and as the name suggests, it forms packages from roll around a central mandrel that incorporates a product fill tube and allows the forming, filling, and sealing of packages of almost all types of products from solids to liquids like milk. Mostly these may be multiple-head machines to produce small packets of sauces for fast-food restaurant use. With many types of food products, the product handling must be done accurately as the filled pouches are filled to avoid infecting the seal surfaces. Additional types of machinery peculiar to the food industry will insert slices of meat products into folded or vacuum formed film and seal it, or will even extrude a cylinder of cheese food into a tube of plastic film and then flatten and seal the product to create the wrapped processed cheese slices.

Specialty machinery may be necessary for particularly sensitive materials such as fruit juices and meats that may discolour with the slightest gain of dissolved metal. In the most extreme cases like processing of juices and pharmaceutical or cosmetic products, special surfaces must be constructed for contact with the product using materials such as borosilicate glass, Teflon, or other materials, as appropriate. For some products the film may first be fed through a sterilizing chemical bath and dryer prior to use in the packaging system.

The film approaches the back of a long hollow conical tube, and when the centre of the plastic is near the tube, the outer edges of the film form flaps that wrap around the conical tube. The film is pulled downward around the outside of the tube and a vertical heatsealing bar clamps onto the edges of the film, bonding the film by melting the seam edges together.

To start the bagging process, a horizontal sealing bar clamps across the bottom edge of the tube, bonding the film together, and cutting off any film below. The sealed tube end is then lowered onto a precision weighing table and the product to be bagged is dispensed through the long conical tube in the center of the bag.

When the tare weight of the product-filled bag is reached, filling stops, and the horizontal sealing bar seals the top of the bag, and simultaneously forms the bottom of the next bag above. This bag is then cut off from the tube and is now a sealed package, ready to advance onward into the product boxing and shipping processes.

Packaging in Cans and Tubes

Cans made of tin plate are used for sweetened condensed milk, evaporated milk, canned fruits, canned beans etc. The cans open at the back are loaded into the machine which fills the product in to the cans. Then the lids are applied and closed by either simple or double seaming depending on the product. Aerosol cans are used for the packaging of whipped cream. In these, the product is filled with a propellant which is under pressure.

Flat bottomed collapsible tubes made from aluminium are used for packing cream. After the product is filled from the back of the tube, the head space in the tube is filled with inert gas and the tube is then closed by folding. Polyethylene collapsible tubes are presently used for packing cheese spread/cheese food and ketchups.

Shrink Wrapping Machines

Shrink wrapping includes packing of one or several articles with a thermoplastic film which when subjected to heat shrinks and form a tight wrap around the object. Shrink wrap or shrink film, is a material made up of polymer plastic film. When heat is applied to this material it shrinks tightly over whatever it is covering. Heat can be applied with a hand held heat gun or the package can pass through a heat tunnel on a conveyor.

Shrink wrap is usually used as an overwrap on many types of packaging, including cartons, boxes, beverage cans and pallet loads. A variety of products may be enclosed in shrink wrap to stabilize the products, unitize them, keep them clean, add a degree of tamper resistance, etc. It can be primary covering for foods like cheese and Paneer.

Shrink wrap film is available in a variety of thicknesses, clarities, strengths and shrink ratios. The two primary films are either crosslinked, or non-crosslinked. Major shrink films include PVC and several other compositions like LDPE, LLDPE, PP, EVA etc. Co-extrusions and laminations are available for specific mechanical and barrier properties for shrink wrapping food.

In shrink-wrap machine a loose plastic film pouch is made. The product is placed in this pack which passes through a heated tunnel in which the film shrinks and adheres closely to the product. The film is commonly heated by hot air, infrared rays or hot water. Shrink wrapping is also used to hold together several singly wrapped products or small packages in a multiple unit package.

Advantages of shrink wrap packaging are:

- All types of items of regular/irregular shapes and sizes can be shrink wrapped.
- Small items can be utilized and stacked one on top of the other.
- Requires minimum packaging material and operation.
- Simple operation.
- Easy stacking.
- Enable unit packaging or packaging in groups.

Robotic Machines

Robotic operations are most effective in operations that are repetitive and predictable but require either control of heavy goods, as the case with palletizers, or with fast, accurate pick-and-place operations that needs consistent, accurate results at speediness that human operators would have difficulty maintaining. The second application is more usual in food packaging operations, and one of the first combination of robotics in the food packaging industry was in the high-speed filling of boxes of assorted chocolates where number of different products had to be accurately placed in a moving container.

Many robotic fittings mimic the human arm, or involve an articulated manipulator set on a revolving base, called delta robots. They have been successfully incorporated into many packaging operations. Originally developed by Raymond Clavel at Ecole Polytechnique Federale de Lausanne, these can be made with high-speed actuators and lightweight carbon fibre elements to provide extremely accurate high-speed operation for the management of small light objects. They have been implemented in many applications ranging from machine parts to bakeries and confectionary manufacturing.

Pre-Closure Treatment and Closure Applications

A pre-closure treatment like headspace gas flushing is necessary for many closures and seals before they may be applied. This may require a remarkably complex system of nozzles and diffusers to avoid blowing the product out of the package or contaminating the seal. Vacuum may be applied, either alone or as a part of the gas-flush process as in a draw-redraw flushing where first air is withdrawn, then replaced with gas, and the process is repeated to ensure saturation of the flush gas. This procedure is common in products with high interstitial volume, such as bagged salads.

Case Packers

Although case packers are assumed to be a distinct type of machinery, it is not strange to have case packers that go home at the end of their shift. Several operations still depend on manual labour for some level of case aggregation, filling, and packing. Even though tedious, this can be helped by accumulator tables where product can be lifted in case quantities or inverted cases turned over for the final grouping. Bigger operations may automate many of the tasks, with machines lifting case quantities into shipping containers and sealing them. At this point, particularly for small consumer items and food, the handling of the filled case ask for some attention to the ergonomics of the material movement, because the filled case may be heavy enough to cause health or safety problems for the personnel involved.

Palletisers

Manual palletisation may be used much as with case packing, but owing to the risk of repetitive motion injuries and the chance of dropped cases, there is often mechanical assistance used in the high-speed accumulation of pallet loads of finished goods. Palletisation is one of the important areas where large-scale automation and robotics were incorporated into packaging operations because of the blend of load weight, precision placement, and speed of operation. As heavy-lift robot arms become more capable, they may be desired for ease of setup and flexibility of mixed product loads.

If mechanical support is used in palletizing or handling, the case itself must be designed with the material handling system in mind. For example, vacuum-lifting structures in an automatic palletizer may put unexpected tensile stress on box flaps that will pull them open if the box-closing adhesive or tape is inadequate.

Stretch Wrapping

Stretch wrap or stretch film is a highly stretchable plastic film that is wrapped around items. In contrast, shrink wrap is applied loosely around an item and shrinks tightly with heat. It is frequently used to unitize pallet loads but also may be used for bundling smaller items. Types of stretch film include bundling stretch film, hand stretch film, extended core stretch film, machine stretch film and static dissipative film.

The most common stretch wrap material is LLDPE, which is produced by copolymerization of ethylene with alpha-olefins, the most common of which are butene, hexene and octene. Other types of polyethylene and PVC can also be used. Many films have about 500% stretch at break but are only stretched to about 100 – 300% in use. Once stretched, the elastic recovery is used to keep the load tight. Other properties such as break strength, cling, clarity, tear resistance, static discharge, etc. are also to be considered. In pallet unitizing, stretch wrap can have several functions like:

- Improved stability of products or packages, forming a unit load.
- More efficient handling and storage of unit loads.
- Some degree of dust and moisture protection.

- Some degree of tamper resistance and resistance to package pilferage.
- Stretch wrapping is the most cost-effective way to keep loads secured on a pallet.

Stretch wrapping can be applied manually with small rolls of film. Dispensers are also available for larger rolls. Machinery is available to automate the operation. This controls the amount of material used, controls the stretch, and controls the application pattern. It is important not to apply too much tension or too many layers: the stress can damage the vertical edges of the boxes and significantly reduce stacking strength.

Speciality Packaging Machines

Inert Gas Packing

Inert gas packing using nitrogen, carbon dioxide or a mixture of the two is done by passing the gasses around the product prior to sealing for example, Cheese. Whole milk powder packed in tins with a pin hole is evacuated under vacuum. The vacuum is then broken by the inert gas (usually N_2) and the pin hole is sealed immediately.

Vacuum Packaging

Vacuum packaging is done for products like cheese blocks, meat products etc, where there is problem of microorganisms growing on the surface. The product is placed in a plastic pouch and placed in the vacuum packaging machine for the creation of vacuum in the pack and subsequent sealing takes place in the machine itself.

Materials used in Food Packaging

Selecting the suitable material for packaging a certain type of food depends on the functions that the package is supposed to fulfil. These functions include shielding the foods against moisture, temperature variations, oxygen, light, and biological microorganisms. Also, damage protection, permeability, food identification, and chemical and optical properties play a significant role in the material selection. Conventional food packaging materials currently used vary between metals, paper, glass, and plastics.

Metals

There are various forms of metal food packaging, such as cans, tubes, containers, films, caps and closures. Cans are generally made of aluminium or steel, and they are the most commonly used metal packages of food and beverages. They are highly recyclable and are usually coated with a layer of organic material to prevent any interaction between the food and the metal.

Aluminium

Aluminium is generally used for beverage cans, foils, tubes, trays, pouches, and coffee capsules. It has good resistance to temperature fluctuations and acts as an excellent gas barrier, which extends the food's shelf-life. It has outstanding malleability and formability and can be easily embossed. It is relatively harmless, lightweight, and can be recycled indefinitely. Alloying elements, such as magnesium and manganese, are sometimes added to aluminium to enhance its strength. Aluminium can be used in rigid, flexible, and semi-flexible packaging. It helps maintain the freshness and aroma of the foods, and is good for protection from radiation, oxygen, moisture, oils, and microorganisms. Soft-drinks, seafood, and pet-food are commonly enclosed by aluminium packages. Some aluminium grades used in food packaging include AA 3003 (O, H22, H24), AA 8006, AA 8011, AA 8079, and AA 1235.

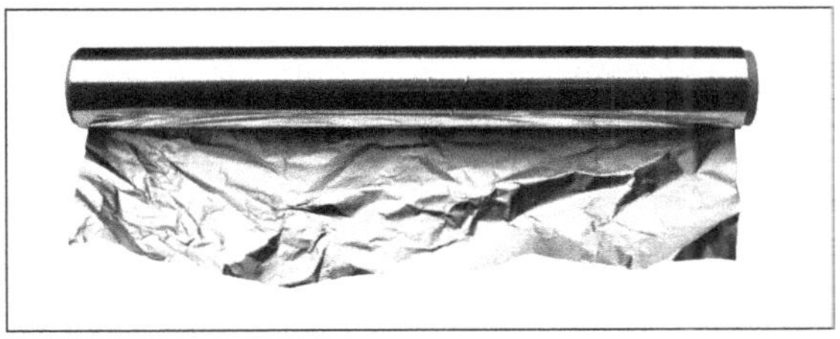

Steel

Steel is used for cans, containers, caps, and closures. Organic coatings are also required to resist corrosion. Steel cans are fabricated from tinplate, which is tin-coated steel, or from electrolytic chromium coated steel (ECCS), also known as tin-free steel. Steel, being a permanent material, can be recycled open-endedly while retaining its quality.

Tinplate is a remarkable barrier to gases, water vapour, light, and odours. It has good ductility and formability, and is convenient for sterile products for it can undergo heat treatment and hermetic sealing. It is light with significant mechanical strength and is suitable for expressive decoration. Common applications of tinplate include drink cans, processed foods, and powdered foods.

Tin-free steel also has good strength and formability and is slightly cheaper than tinplate. The chrome/chrome oxide in ECCS renders it a good material for coating adhesion, such as lacquers and paints. It has good resistance to heat and black sulphide stain, which makes convenient for making fish cans. Applications of tin-free steel include food cans, trays, bottle caps, can ends, and closures.

Paper

Paper is one of the oldest packaging materials, dating back to the 17th century. Paper and paperboard are mostly used for packaging dry foods. Upon coating or waxing, their applications extend to packaging and serving of wet and fatty foods. They are commonly used in corrugated boxes, milk cartons, folding cartons, paper plates and cups, bags and sacks, and wrapping paper.

Paper is used for temporary food containment and protection due to its high permeability and inability to be sealed with heat. When used as primary packaging, waxes, resins, and lacquers are used as coatings and laminates to enhance the paper's protective and functional properties. Depending on its method of production and packaging purpose, paper can be found as Kraft paper, sulphite paper, greaseproof paper, Glassine, or parchment paper. Kraft paper is the strongest form of paper, and is used in packaging flour, sugar, and dried fruits. Sulphite paper is relatively weaker and lighter, and is used to wrap biscuits and sweets. Greaseproof paper and Glassine contain densely packed cellulose fibres. Glassine being further hydrated that improve the paper's oil resistance, thus making it suitable to package snacks, biscuits, fast foods, and greasy foods. Parchment paper is acid-treated paper, which renders it impermeable to fluids, but not to air and vapour. It is used in packaging butter and lard.

Paperboard is a relatively thicker and heavier material than paper. It is widely used as secondary packaging that is not in direct contact with the food. Boxes, trays, and cartons used for shipping are the common usages of paperboard. Types of paperboard vary between white board, solid board, chipboard, and fibreboard. White board is the only paperboard advised for primary packaging. Solid board is a strong and durable paperboard, used to package milk, fruit juices and soft drinks. Chipboard is the cheapest form of paperboard, made of recycled paper, and is used as outer layers of food cartons, such as cereals and tea. Fibreboard is used to ship bulk food due to its strength and resistance to impact scratching and crushing damage.

Another form of paper is paper laminates, which are coated or uncoated paper based on Kraft and sulphite tissues. They have enhanced barrier properties, and are used in packaging soups, spices, and herbs.

Glass

Glass is another permanent packaging material that has been used for millennia. The earliest evidence of glass making was around 7000 B.C. yet, glassblowing of bottles was discovered by the Romans in 50 B.C. in the area of modern day Lebanon. Glass is well-known for being amid the most reliable and least toxic materials for packaging foods and drinks. Its advantages include imperviousness, inertness, strength, hygiene, resistance to tampering, quality colour, design, decoration potential, transparency, chemical propriety, microwaveability and heat treatability.

There are two types of glass packaging most widely used for foods and drinks: narrow-neck bottles and wide-opening jars and pots. Glass bottles are commonly used for alcoholic drinks, soft drinks, and potable water. Foods packed in glass containers range from coffee, to dairy products, spices, spreads, syrups, processed vegetables and fruits, and meat and fish products. With the rise of popularity and usage of other packaging materials, such as metals and plastics, high-value products are preferred to be packed in glass containers due to their high-quality image and aroma preservation characteristics.

Glass is basically made from soda, lime, and silica in addition to other elements depending on the desired characteristics and is manufactured via melting and container forming. Surface treatment, heat treatment and annealing are subsequent steps in the production process. It is reusable and infinitely recyclable through crushing, melting, and reforming without loss of quality.

Plastics

Plastics are the most common and most wide-ranging materials used for food packaging. Some of their widespread uses are bottles, trays, bags, foils, cups, pots, pouches, and bowls. The volume of plastic allocated to food packaging amounts for around 40% of plastics. The convenience and widespread use of plastic in food packaging is owed to its low cost, ease of processability, formability, chemical resistance, lightweight, and a variety of physical properties. However, plastic suffers from permeability to gas, vapour, and light.

Plastics can be classified into two main categories: Thermosets and thermoplastics. Thermosets are polymers that irrevocably solidify upon heat and are non-reformable, which makes them unsuitable for food packaging. Thermoplastics, on the other hand, soften when heated and are able to retain their initial conditions at room temperature. This renders them perfect for food packaging. Besides, despite certain functional restrictions, thermoplastics are recyclable via melting, reproduction and reuse as new products.

Despite the health and safety a concern regarding residual components from plastic, plastic use continues to grow compared to the aforementioned conventional materials because of its inexpensiveness, thermosealability, microwaveability, and ease of fabrication into countless shapes and sizes.

Of the plastics used for food packaging, polyolefins and polyester are the most common materials. Other materials include polyvinyl chloride, polyvinylidene chloride, polystyrene, polyamide, and ethylene vinyl alcohol.

Polyethylene and polypropylene are materials from the polyolefin category. These two materials are extensively used due to their light weight, malleability, strength, stability, processability, reusability and resistance to chemicals and moisture. Milk, juice, and water bottles, grocery, retail, and garbage bags, and bread and frozen food bags are some of the uses of polyethylene. Polypropylene is used when heat resistance is needed. Yogurt containers and margarine tubs are applications of polypropylene.

The most commonly used polyester in food packaging is polyethylene terephthalate, more commonly known as PETE. PETE is a resistor of heat, oils, solvents, and acids. It has good ductility, strength, and hardness. Its advantageous properties also include lightness, impermeability to gas, transparency, and resistance to breaking. It is mostly used in bottles, tubs, jars, trays, blisters, bags, and wrappers for snacks.

Future of Food Packaging Materials

Innovations and advancements in material science have given food packaging a positive future in terms of efficiency and environmental impacts.

Smart packaging (or Intelligent packaging) has been gaining publicity lately, with an expected market of $26.7 billion by 2024. Smart packaging represents packaging systems carrying sensors that help prolong the shelf life of foods, reveal information on freshness and quality, and enhance the safety of the product and consumer.

Another technology in packaging systems is active packaging, which is the integration of additives with the packages to improve food shelf-life and quality. Nanotechnology, as well, has been able to penetrate the food packaging industry. It helps augment the thermal and mechanical properties of food packages. Nanosensors embedded within the packaging systems can also help monitor, identify, and warn about the safety and quality of the foods. Although the effect of nanoparticles on human health is yet to be well understood, nanotechnology displays a promising future.

When it comes to waste management, packaging waste has occupied a large portion of municipal solid waste (MSW), resulting in a rise of environmental concerns. Biodegradable polymers have risen as alternatives to traditional plastics in food packaging. These polymers decompose into CO_2, water, inorganic compounds, and biomass. The types and applications of some biodegradable polymers are listed in Table.

Table: Different categories, types, and applications of biodegradable polymers.

Biodegradable Polymers		
Category	Polymer Type	Application
Natural Polymers	Starch	Wraps, cups, plates, trays, disposable dishes
	Cellulose	Films, fresh food packaging, baked goods packaging
	Chitosan	Films for fresh fruits and vegetables
	Protein	Films, Coating on paper
Synthetic Biopolymers	Polyhydroxyalkanoates (PHA)	Cheese coatings, fast food plates, bottles, containers, sheets
	Bacterial cellulose (BC)	Films
Non-natural Polymers	Poly(vinyl alcohol) (PVA)	Water-soluble films
	Polylactide (PLA)	Paper coatings, films
	Polyglycolide (PGA)	Protective layer in multilayer packages, such as PETE bottles, cups, dishes, trays
	Poly(e-caprolactone) (PCL)	Foam trays, loose fill, film bags
	Poly(butylene succinate) (PBS)	Films, bags
	Poly(trimethylene terephthalate) (PTT)	Films
	Poly(butylene adipate-co-terephthalate) (PBAT)	Films, blown bottles, disposable food bags, trays

Aseptic Packaging of Food

Heat sterilization is the unit operation in which foods are heated at a sufficiently high temperature for a sufficiently long time to destroy microbial and enzymatic activity present in the raw food. As a result, sterilized foods have a shelf life of 6 to 12 months or more. The most common method of sterilizing solid and viscous food products is in-container sterilization, e.g., canning; however, the main disadvantages associated with in-container sterilization of food products are:

- The low rate of heat penetration to the thermal center of the container.
- Long processing times to achieve the desired sterility.
- Damage to the nutritional and sensory characteristics of the product.
- Low productivity.
- High energy costs.

To overcome the constraints of in-container sterilization, products can be sterilized at higher temperatures for a shorter time prior to filling into pre-sterilized containers under sterile conditions. This forms the basis of ultra-high-temperature (UHT) processing and aseptic packaging, which has been defined as "the independent sterilization of food and packaging and assembly under sterile conditions" or Aseptic packaging can be defined as the filling of a commercially sterile product into a sterile container under aseptic conditions and hermetically sealing the containers so that re-infection is prevented. This results in a product, which is shelf-stable at ambient temperature conditions or Aseptic packaging refers to the filling of a cold, commercially sterile product under sterile conditions into a pre-sterilized container and closure under sterile conditions to form a seal that effectively excludes microorganisms.

Aseptic processing was developed in the early 1940s and has been used effectively in Europe and Japan for over three decades. It has been rapidly gaining reputation as a thermal processing technique in North America since the use of hydrogen peroxide was approved for sterilizing of packaging materials in 1981. The process has long been effectively employed to sterilize a wide range of liquid products, e.g., milk, fruit juices and concentrates, cream, yogurt, salad dressing, egg, and ice cream. Current developments in high-barrier plastic packaging materials and aseptic processing and filling technology have resulted in the process being expanded to sterilize both acid-low and acid-viscous and semisolid foods that contain discrete particles, such as tomato paste, baby foods, fruit pulp etc.

Scope of Aseptic Packaging

There are number of limitations and disadvantages during actual application of this technology. However, we can't ignore the benefits over various lacunas of the process. Thus, it can be concluded that aseptic packaging of sterile/non sterile food and food products is the most significant innovation in the field of food science and technology and there is a big scope in this area.

Major Categories of Aseptic Packaging Systems

- Can system: It includes hermetically sealed cans.
- Bottle systems: Glass containers and plastics bottles fall into this category. The bottles can further be divided into; a) Non-sterile bottles; b) Sterile blown bottles; c) Single station blowing, filling and sealing.
- Sachet and pouch systems: This system is classified into Form-fill-seal systems and Lay flat tubing.
- Cup systems: The aseptic packaging of food into cups can be into; Pre-formed plastic cups and Form-fill and seal cups.
- Carton systems: This type of aseptic packaging system includes Form-fill-seal cartons and Prefabricated cartons.
- Bulk packaging systems: This type of system classified into Metal drum and Bag-in-box Packaging Lines for Aseptic Processing.

There are five basic types of aseptic packaging lines as given below:

- Film and Seal: Pre-formed containers made up of thermoformed plastic, glass or metal are sterilized, filled in aseptic environment and sealed.

- Form, Fill and Seal: Roll of material is sterilized, formed in sterile environment, filled and sealed. e.g. tetra packs.
- Erect, Fill and seal: Using knocked, down blanks, erected, sterilized, filled sealed. e.g. Gable-top cartons, Cambri-block.
- Thermoform, Fill: Sealed roll stock sterilized, thermoformed, filled, sealed aseptically. e.g. Creamers, plastic soup cans.
- Blow, Mold, Fill and Seal: Different package forms used in Aseptic UHT processing are cans, paperboards/plastic/foil/plastic laminates/flexible pouches, thermoformed plastic containers, bag in box, bulk totes.

Pre-Requisite Conditions for Aseptic Packaging

- It should contain the product.
- It should prevent physical damage to packaged product.
- It should run effortlessly on filling lines.
- It should survive packaging processes.
- It should be easy to handle throughout distribution process.
- It should avoid dirt and other contamination.
- It should be able to protect the product from odours and taints.
- It should be resistant to rodent attack.
- It should be able to stop insect infestation.
- It should be biologically safe i.e. non-toxic.
- It should be compatible to foodstuff.
- It should maintain sterility of product.
- It should prevent entry of microorganisms.
- It should show evidence of tampering.
- It should control moisture loss or gain.
- It should offer a barrier to oxygen.
- It should be protective against the light.
- It should maintain gas atmospheres, i.e.CO_2/N_2.
- It should communicate all the information regarding product and manufacturer.
- It should have good sales appeal.

- It should be easy to open and handle.
- It should be cost efficient.

The above given pack criteria are separated into seven areas, mainly as follows:

- Product Containment: The need to contain the product in the sense that liquids or powders do not leak out.
- Physical Protection: This is required when dealing with fragile foods like eggs or snack foods, but minor impacts on fresh fruits, for example, will release enzymes and lead to browning and softening. Equally important is the adverse effects on sales of damaged packages themselves-even though the product is in good condition.
- Food Safety: The need to ensure that the aseptically packed food retains its sterility, through a package that prevents adventitious contamination by microorganisms. Tamper evidence is also a desirable requirement. The other aspect of food safety is the avoidance of long-term chronic effects from the food packaging materials themselves.
- Shelf-Life: For dried foods moisture gain is a major factor in determining shelf-life. Atmospheric oxidation, often catalyzed by light, is more critical for aseptically packed foods such as milk, fruit juices, or cream soups. Hence a good oxygen and light barrier, as provided by tinplate or aluminum foil, is needed to ensure maximum shelf life for aseptically packed products.
- Communication of Information: The package should need to tell the purchaser what food is inside it and whose product it is. Apart from this, more information should be passed on to the customer, such as net weight, list of ingredients, batch number, useby date, nutritional information etc.
- Sale-Appeal: The package must look attractive and "catch the eye" of prospective purchasers, and it should also be easy to open and dispense the product.
- Cost-Effectiveness: Value for money in packaging is more important than looking for the lowest price. A cheap but dimensionally variable container could cause more damage during production or an increase of "leakers" in the market place thereby affects the sale of the product.

A currently popular packaging material used for aseptic packaging is a paperboard/foil/plastic laminate made by Tetra-Pak. This laminated structure consists of as many as six layers of materials viz. polypropylene, surlyn, aluminum foil, polyethylene, paperboard, and polyethylene as the innermost layer.

Other variations that can also be used for thermal sterilization include laminates consisting of saran, ethyl vinyl alcohol (EVOH), polyethylene, and polystyrene or metalized polyester consisting of vinyl ethyl acetate, nylon, foil, and polyethylene. Aluminum foil is the most commonly used barrier material, with polypropylene or polyolefin (type of polyethylene) being the universal heat-sealing and food contact surfaces. When foil is used, it needs to be protected against mechanical harm, which is usually provided by paperboard. All of these composite packages yield the desirable moisture, oxygen, light and microbial barrier properties, strength and heat stability needed for successful aseptic package.

Aseptic Tank

The aseptic tank is used for intermediate storage of UHT treated food products. It can be used in different ways in UHT lines, depending on plant design and the capacities of the various units in the process and packaging lines.

- If one of the packaging machines incidentally stops, the aseptic tank can take care of the surplus product during the stoppage.
- Simultaneous packaging of two products.

The aseptic tank is first filled with one product, sufficient to last for a full shift of packaging. Then the UHT plant is switched over to another product which is packed directly in the line of packaging machines. One or more aseptic tanks included in the production line offer flexibility in production planning.

Direct packaging from a UHT plant requires recirculation of a minimum extra volume of 300 litres per hour to maintain a constant pressure to the filling machines. Products which are sensitive to overtreatment cannot tolerate this and the required overcapacity must then be fed from an aseptic tank. The optimum arrangement must thus be decided for each individual process with UHT plants, aseptic tanks and aseptic packaging machines.

Role of Microprocessors in Aseptic Packaging

Microprocessors and microcomputers are first used in packaging machines in 1973. Microprocessor-based equipment controls were first used in 1977. Microprocessor-controlled packaging machinery was first commercially used in 1978. Microprocessor-based aseptic packaging has capability of monitoring one or more process variables simultaneously.

Main operations that are taken care by microprocessors are:

- Feeding of film to the machine.
- Converting of film into required shape of specific dimensions.
- Filling the product with specific volume of product.
- Heat sealing.
- Collecting up of specific no of packs and shrink wrapping them into one single pack.

All these operations are taken care by Microprocessors.

Sterilization of Packaging Material

Whatever the choice of packaging material used, it must be pre-sterilized prior to filling. Sterilization of the packaging material should not impair that material. Methods commonly used for sterilization of packaging materials or packages include steam (saturated or superheated), hot air, hydrogen peroxide, ultraviolet light, irradiation, or the heat generated during co-extrusion of certain films.

Since aseptic packaging systems are complex, there is considerable scope for packaging faults to occur, which will lead to spoiled products. Thus, following precautions are to be taken.

- Packages should be inspected regularly to ensure that they are airtight, again focusing upon those more critical parts of the process, such as start-up, shutdown, product changeovers and, for carton systems, reel splices and paper splices.
- Pipes, storage tank, and surfaces of the packaging machine come into contact with the sterilized product have to be sterilized.
- Sterilization procedures should be verified.
- The seal integrity of the package should be monitored as well as the overall microbial quality of packaging material itself.
- Care should be taken to minimize contamination during subsequent handling. All these could result in an increase in spoilage rate.
- Rinsing, cleaning and disinfecting procedures are also very important, especially the removal of fouling deposits, which may provide a breeding ground for the growth of micro-organisms, especially thermofiles.

Quality Assurance Aspects of Aseptic Packaging

Aseptic packaging has to be meticulously checked. Not only must the packaged product be examined, but so must all preceding steps, as well as the operators, which are potential carriers of pathogens. If just one bacterium reaches the product, and that bacterium is pathogenic and can proliferate (for example, Staphylococcus aureus), the result could be disastrous. In addition to regular sampling during production, further samples should be taken at the times or in situations known to be associated with an increased risk of contamination. It is advisable to incubate these samples long enough, in most cases from 5 to 7 days at 30 °C to allow sub-lethally damaged bacteria also to grow to detectable counts. The products should only be delivered if the result of the shelf-life test is satisfactory.

Active and Intelligent Packaging

Active packaging can be defined as "Packaging in which supplementary constituents have been deliberately added in or on either the packaging material of the package headspace to improve the performance of the package system". In simple terms, Active packaging is an extension of the protection purpose of a package and is commonly used to protect against oxygen and moisture. It permits packages to interact with food and environment. Thus, play a dynamic role in food preservation. Developments in active packaging have led to advances in many areas, like delayed oxidation and controlled respiration rate, microbial growth and moisture migration. Other active packaging technologies include carbon dioxide absorbers/emitters, odour absorbers, ethylene removers, and aroma emitters.

On the other hand, intelligent packing can be defined as "Packaging that has an external or internal indicator to provide information about characteristics of the history of the package and/or the quality of the food packed". Intelligent packaging is an extension of the communication purpose of conventional packaging and provides information to the consumer based on its ability to sense, detect, or record external or internal changes in the product's environment. Eg: time-temperature indicators, ripeness indicators, biosensors, and radio frequency identification.

- Active packaging changes the condition of the packed food to extend shelf-life or to improve safety or sensory properties, while maintaining the quality of the packaged food.
- Intelligent packaging systems monitor the condition of packaged foods to give information about the quality of the packaged food during transport and storage.

Active Food Packaging Techniques

Active packaging technologies have started to receive a great deal of attention since the last decade. The market for active packaging films was about $50 million worldwide in 2004, and this market is growing rapidly.

Active packaging techniques for preservation and improving quality and safety of foods can be divided into three categories: absorbers or scavengers, releasing systems and other systems. Absorbing (scavenging) systems remove undesired compounds such as oxygen, carbon dioxide, ethylene, excessive water, and other specific compounds. Releasing systems actively add or release compounds to the packaged food or into the head-space of the package such as carbon dioxide, antioxidants and preservatives. Other systems may have miscellaneous tasks, such as self-heating, self-cooling and preservation.

Depending on the physical form of active packaging systems, absorbers and releasers can be a sachet, label or film type. Sachets are placed freely in the head-space of the package. Labels are attached into the lid of the package. Direct contact with food should be avoided because it impairs the function of the system and, on the other hand, may cause migration problems.

Oxygen Scavengers

High levels of oxygen present in food packages may facilitate microbial growth, off-flavours and off-odours development, colour change, and nutritional losses, thus causing significant reductions in the shelf-life of foods. Thus, the control of oxygen levels in packages is important to limit the rate of such deteriorative and spoilage reactions in foods. Oxygen absorbing systems provide an alternative to vacuum and gas flushing and improve product quality and shelf-life. Also, they are economically viable in reducing packaging costs and increasing profitability.

Typical oxygen absorbing systems are based on the oxidation of iron powder by chemical means or scavenging of oxygen by the use of enzymes. In the former, iron kept in a sachet is oxidized to iron oxide. The sachet material is highly permeable to oxygen and in some cases, to water vapour, for the sachet to be effective. This well-known oxygen scavenging system was first developed and introduced to the food packaging market by the Mitsubishi Gas Chemical Company, known as Ageless. The type and amount of absorbent that needs to be used in a sachet is determined by the initial oxygen level in the package, the amount of dissolved oxygen present in food, permeability of the packaging material, nature (size, shape, weight, etc.) of food and the water activity of the food. These iron-based oxygen absorbing systems have the ability to scavenge oxygen in many foods, including high, intermediate, or low moisture foods, and foods containing lipids. They can also work at refrigerated and frozen storage conditions, and can be used as effective oxygen scavengers with microwaveable food products.

In enzymatic oxygen scavenging systems, an enzyme reacts with a substrate to scavenge oxygen. These systems are more expensive than iron-based systems, due to the cost of enzymes used for the oxygen scavenging purpose. Enzymatic oxidation systems are also usually very sensitive to temperature, pH, water activity and solvent/substrate present in the sachet, thus limiting the widespread use of these enzyme-based systems.

Oxygen scavenging sachets are not appropriate for liquid foods, because the direct contact of the liquid with the sachet usually causes the leakage of sachet contents. Also, sachets may cause accidental consumption with the food or may be ingested by children. Oxygen scavenging sachets sold in the US are required to be labelled "Do not eat," for safety reasons and regulatory purposes mandated by the Food and Drug Administration. Although sachets can be concealed using secondary packages, but this practice increases packaging costs.

The incorporation of scavengers in packaging films is a good way of resolving sachet-related problems. Scavengers may either be inserted into a solid, dispersed in the plastic, or inserted into various layers of the package, including adhesive, lacquer, or enamel layers. Multi-layer oxygen scavengers more effectively absorb oxygen than single layer scavenging systems.

In the structure of a typical multi-layer oxygen scavenging system, an oxygen absorbing substance imbedded in a layer very permeable to oxygen is used to absorb oxygen present in the package head space. Oxygen entry from the outside environment to the oxygen absorbing layer is limited by a barrier layer, which is highly impermeable to oxygen. An inner layer, or a control layer, next to the oxygen absorbing layer is used to minimize any migration of the oxygen absorbing substance into the food.

Scavengers in the form of films allow the absorption of oxygen from all surfaces of the food that are in contact with the film. The rate of oxygen absorption by films changes, depending on the area and thickness of the film.

The Cryovac Corporation has developed a polymer based film that can overcome the negative effects of iron-based oxygen scavenging films. This film uses innovative technology that is invisible to consumers because the scavenging component is co-extruded as a layer of the package. Since oxygen absorbing material is virtually invisible in the films, these films do not alter the look of the package. They also offer shoppers a clear view of the product packed. It is equally effective with wet and dry products, because its scavenging action is initiated "on demand" on the processor's packaging line by an ultraviolet-light triggering process.

Carbon Dioxide Scavengers and Emitters

High levels of carbon dioxide usually play a beneficial role in retarding microbial growth on meat and poultry surfaces and in delaying the respiration rate of fruits and vegetables. For instances where the package has a high permeability to carbon dioxide, a carbon dioxide emitting system may be necessary to reduce the rate of respiration and suppress microbial growth. The use of a dual function system consisting of an oxygen scavenger and a carbon dioxide emitter is the usual practice for increasing the shelf life of highly perishable foods.

On the contrary, dissolved carbon dioxide formed after the roasting of coffee may cause the package to burst, if the roasted coffee is packed in a can or aluminium foil pouch. This released carbon dioxide from freshly roasted coffee can be scavenged through the use of a carbon dioxide scavenger. A carbon dioxide absorbing sachet is developed, which is composed of a porous envelope containing calcium oxide and a hydrating agent, such as silica gel, on which water is adsorbed. In this system, water reacts with calcium oxide and produces calcium hydroxide, which then reacts with carbon dioxide to form calcium carbonate.

Moisture Scavengers

If the package has a low permeability to water vapour, water accumulation inside the package is more pronounced. The excess water development inside a food package usually occurs due to the respiration of fresh produce, temperature fluctuations in high equilibrium relative humidity food packages or the drip of tissue fluid from cut meats, poultry, and produce. An effective way of controlling excess water accumulation in a food package that has a high barrier to water vapour is to use a moisture or water vapour scavenger, like silica gel, molecular sieves, natural clays, calcium oxide, calcium chloride and modified starch, or other moisture absorbing substances. Silica gel is the most widely used desiccant because it is nontoxic and non-corrosive.

A blanket or a pad of desiccant is usually wrapped around the food to be preserved (meat, poultry or fish) to absorb water. Another approach to absorb moisture from these foods is to use a superabsorbent polymeric laminate film that has a moisture absorbent layer that is formed from a polyester graft copolymer and a resin, consisting of a polyurethane resin, an acrylic resin and vinyl resin. Moisture absorbing systems in sachet are usually used to maintain low levels of moisture in dried food packages, like chips, nuts, spices, biscuits, crackers, milk powders and instant coffee. The sachets like Desi Pak, Desi View, Sorb-It and 2-in-1 from United Desiccants, USA, MiniPax, StripPax, Natrasorb, and the moisture absorbing label Desimax from Multisorb Technologies, USA are the most common moisture absorbing systems used to absorb and/or control moisture for water sensitive packaged foods in USA.

Ethylene Absorbers

Ethylene is a hormone that accelerates ripening and senescence by increasing the respiration rate of climacteric fruits and vegetables, thereby decreasing shelf-life. Hence, the removal of ethylene gas from the package headspace slows senescence and improves shelf-life.

The most well-known, inexpensive, and extensively used ethylene absorbing system consists of potassium permanganate inserted in silica. The silica absorbs ethylene, and potassium permanganate oxidizes it to ethylene glycol. Silica is kept in a sachet permeable to ethylene, or it can be incorporated into a packaging film. Potassium permanganate is not integrated into food contact surfaces of packaging films due to its toxicity. The substrate surface area and the amount of potassium permanganate affect the performance of these systems.

Another system is based on infusing zeolite with potassium permanganate, and then coating the infused zeolite with a quaternary ammonium cation. This system is not only capable of absorbing ethylene from the medium, but also other organic compounds, such as benzene, toluene, and xylene. Ethylene scavengers have been proven to be effective in the storage of packaged fruits, including kiwifruit, bananas, avocados, and persimmons. Ethylene scavengers are commercially available as Evert-Fresh (Evert-Fresh Co, USA), Ethylene Control (Ethylene Control Incorporated, USA) and Peakfresh (Peak Fresh Products, Australia).

Flavour Absorbing/Releasing Systems

Food packaging materials like some plastics may interact with food flavours, resulting in loss of flavours, known as flavour scalping. Flavours are usually lost or degraded after processing foods at high temperatures or after packaging. Therefore, there is a need to replace these lost flavours a constituent when scalping occurs. In addition, consumers always like to smell good flavours when they first open a food package. For example, most of the dry instant coffee manufacturers often fill the headspace with distilled volatiles to have fresh coffee fragrance when the package is first opened.

In contrast to flavour releasing systems, flavour absorbers scavenge undesirable flavours, aromas, and odours present in the package headspace. Some orange varieties, such as Navel may develop a bitter flavour when limonin concentration exceeds 12 mg/kg after pasteurization. Internally cellulose acetatebutyrate coated plastic bottles, which act as limonin absorbers, have been shown to considerably reduce bitterness in orange juice thereby increasing shelf-life.

The formation of off-flavours and off-odours in food products are due to (1) the oxidation of fats and oils, leading to the formation of aldehydes; (2) the breakdown of proteins of fish muscle into amines. Although aldehydes and amines can be removed from package headspaces by flavour scavengers, these systems may mask or absorb off-flavours and off odours that are indicative of spoilage.

Intelligent Packaging Techniques

The definition of intelligent packaging in the Actipak project includes indicators to be used for quality control of packed food. They can be so-called external indicators, i.e., indicators which are attached outside the package like time-temperature indicators, and so-called internal indicators which are placed inside the package, either to the head-space of the package or attached into the lid like oxygen indicators for indication of oxygen or package leak, carbon dioxide indicators, microbial growth indicators and pathogen indicators. Various examples of internal and external indicators are shown in the table below:

Indicator	Principle/reagents	Gives information about	Application
Time-temperature indicators (external)	Mechanical, Chemical, Enzymatic	Storage conditions	Foods stored under chilled and frozen conditions.
Oxygen indicators (internal)	Redox dyes, pH dyes Enzymes	Storage conditions Package leak	Foods stored in packages with reduced oxygen concentration.

Carbon dioxide indicator (internal)	Chemical	Storage conditions, Package leak	Modified or controlled atmosphere food packaging.
Microbial growth indicators (internal /external) i.e. freshness indicators	pH dyes, All dyes reacting, with certain metabolites (volatiles or non-volatiles).	Microbial quality of food (i.e. spoilage)	Perishable foods such as meat, fish and poultry.
Pathogen indicators (internal)	Various chemical and immunochemical methods reacting with toxins.	Specific pathogenic bacteria such as Escherichia coli O157.	Perishable foods such as meat, fish and poultry.

Time-Temperature Indicators

Temperature abuse is common during storage, transportation, and distribution these indicators are designed to monitor temperature abuses in a food product's shelf life. Temperature abuse does not only cause quality and nutritional losses, but also may lead to food poisoning and food losses. Many time-temperature indicating systems are based on a colour change or colour development, which is correlated with food quality loss. Nevertheless, time-temperature indicators that have a response in the form of visible mechanical deformation are also available. Time-temperature indicators must satisfy some requirements to be effective as monitoring devices such as:

- They must be easily activated and sensitive.
- They must provide a high degree of accuracy and precision.
- They should have tamper evident characteristics.
- Response should be irreversible, reproducible, and correlated with food quality changes.
- Physical and chemical characteristics of time-temperature indicators should be determined.
- Response should be easily readable and, should not be confusing.

Time-temperature indicators for frozen, refrigerated, modified atmosphere packaged and thermally processed foods are commercially available. Although time-temperature indicators are still in their infancy, continued research and consumer appreciation are expected to play positive roles in the development of more reliable and sophisticated systems.

Freshness Indicators

A freshness indicator indicates directly the quality of the product. The indication of microbiological quality is, for example, based on a reaction between the indicator and the metabolites produced during growth of microorganisms in the product.

The freshness indicators are mainly based on compounds like glucose, organic acids and their salts, ethanol, volatile nitrogen compounds, carbon dioxide and sulphuric compounds. For example, Concentration of D-lactate increases with storage of meat. Thus it can be taken as freshness indicator.

An indicator that would show specifically the spoilage or the lack of freshness of the product, in addition to temperature abuse or package leaks, would be ideal for the quality control of packed products. The number of concepts of package indicators for contamination or freshness detection of food is still very low, however new concepts of freshness indicators are patented and new commercially available products are likely to become available in the near future.

Other Active and Intelligent Packaging Systems

Other active packaging applications that are expected to find increased consideration in the future include antioxidant releasing films, colour containing films, light absorbing/regulating systems, anti-fogging and anti-sticking films, susceptors for microwave heating, gas permeable/breathable films, and insect repellent packages. For example, antioxidant incorporated films can be used to prevent the oxidation of fats and oils that lead to rancidity. They can also be used to prevent the formation of off-odours and off-flavours in foods.

Packaging of Fresh Fruits and Vegetables

Fruit and vegetables play an important role in healthy nutrition and are high on the list of consumer priorities. However the major obstacle of purchasing ready-to-eat fresh-cut fruits and vegetables is their short shelf life, leading to quick degeneration and decomposition of the product and undesirable look and negative palatability. Fruit and vegetables are living products undergoing a ripening and at the end an ageing process, in which the plant tissue is broken down. The products undergo various biological processes, which also continue after the products have been harvested. For example there is wide variation in respiration rates and ethylene production between fruits and vegetables.

Table: Respiration rates of selected commodities.

Commodities	Respiration Rate at 5 °C (mL CO_2 kg^{-1} h^{-1})
Dried fruits and vegetables, nuts	<5
Apple, grape, potato (mature), onion, garlic	5–10
Apricot, carrot, cabbage, cherry, lettuce, peach, plum, pepper, tomato	10–20
Blackberry, cauliflower, lima bean, raspberry, strawberry	20–40
Brussels sprouts, green onion, snap bean	40–60
Asparagus, broccoli, mushroom, pea, spinach, sweet corn	>60

Table: Classification of fruits according to their maximum ethylene production rate.

Commodities	Ethylene production rate (μL/kg. hr at 20 °C)
Citrus fruits, strawberry, artichoke, cauliflower, cherry, asparagus, potato, grape, leafy vegetables, root vegetables	Very low: 0.01-0.1
Pineapple, blueberry, raspberry, blackberry,cranberry, aubergine, persimmon, cucumber, olive, sweet pepper, chili pepper, pumpkin, watermelon	Low: 0.1-1.0

Banana, fig, mango, honeydew melon, mango, tomato	Moderate: 1.0-10.0
Kiwifruit (ripe), apple, avocado, plum, peach, apricot, nektarine	High: 10.0-100.0
Exotic fruit, cherimoya	Very high: >100.0

Conventional Packs

It is essential to minimise physical damage to fresh produce in order to obtain optimal shelf-life. The use of suitable packaging is vital in this respect. The most common form of packaging in this sector is the use of the fibreboard carton; however, for most produce, additional internal packaging, for example tissue paper wraps, trays, cups or pads, is required to reduce damage from abrasion. For very delicate fruits, smaller packs with relatively few layers of fruits are used to reduce compression damage. Moulded trays may be used which physically separate the individual piece of produce. Individual fruits may also be wrapped separately in tissue or waxed paper. This improves the physical protection and also reduces the spread of disease organisms within a pack.

Packaging can be a major item of expense in produce marketing, so the selection of suitable containers for commercial-scale marketing requires careful consideration. Besides providing a uniform-size package to protect the produce, there are other requirements for a container:

- It should be easily transported when empty and occupy less space than when full, e.g. Plastic boxes which nest in each other when empty, collapsible cardboard boxes, fibre or paper or plastic sacks.
- It must be easy to assemble, fill and close either by hand or by use of a simple machine.
- It must provide adequate ventilation for contents during transport and storage.
- Its capacity should be suited to market demands.
- Its dimensions and design must be suited to the available transport in order to load neatly and firmly.
- It must be cost-effective in relation to the market value of the commodity for which used.
- It must be readily available, preferably from more than one supplier.

Owing to perishability, the bulk of fresh fruit and vegetables are still sold unpacked at retail or wholesale level in perforated polyethylene (PE) bags or nets. Long term transport of these materials must remain minimal to avoid brushing and consequent decay of the loosely package product. When packaging is required at the source or when an extended storage life is desired, the packaging film should have a high gas permeability and anti-fog properties. The packaging of fresh vegetables and fruit provides the largest single use of printed PE bags. The pre-packaging of potatoes, carrots, onions, parsnips, beets, radishes etc. for market sales has become an important aspect of food distribution. Because of low water vapour permeability of PE and polypropylene (PP), both films and bags are sometimes perforated to allow the product to 'breathe'. Permeation is the key to extending fresh produce shelf life.

For most commodities light is not an important influence in their post-harvest handling. However green vegetables, in the presence of sufficient light, could consume substantial amounts of CO_2 and produce O_2 through photosynthesis. Shock and vibration leads to damage to produce cells which causes an increase in respiration and may lead to enzymes being released that will cause browning reactions to begin.

Modified Atmosphere Packaging (MAP)

Polymeric films have been used to package fresh products for over 35 years, with a number of benefits, including control of water loss, protection from skin abrasion and reduced contamination of the produce during handling. They also provide a barrier to the spread of decay from one unit to another. These films will also affect the movement of respiratory gases depending on the relative permeability of the film. This can lead to the development of lowered O_2 and raised CO_2 levels within the package and, as with control atmosphere (CA) storage this can reduce the respiration of the produce and potentially extend shelf-life.

A modified atmosphere (MA) can be created within the pack in two ways. Active modification involves the pulling of a slight vacuum within the pack and then replacing the atmosphere with the desired gas mixture. Absorbers (active packaging) of CO_2, O_2 or ethylene may be included within the pack to control the concentration of these gases. In passive modification systems, the atmosphere is attained through the respiration of the commodity within the pack.

The final equilibrium atmosphere will depend on the characteristics of the commodity and the packaging film. Temperature control is extremely important with MAP, as this will influence the gas permeability properties of the film as well as the respiration rate of the product. One of the main drawbacks to MAP is the potential for O_2 levels to fall too low and cause the production of undesirable off-odours caused by fermentation of the tissues.

The MA packaging technique consists of the enclosure of respiring produce in polymeric films in which the gaseous environment is actively or passively altered to slow respiration, reduce moisture loss and decay and/or extend the shelf life of the products. Many of the films used in MAP, singly do not offer all the properties required for a modified atmosphere pack.

To provide packaging films with a wide range of physical properties, many of these individual films are combined through processes like lamination and co-extrusion. There are several groupings in MAP films. Polyethylene is most commonly used to provide a hermetic seal and also as a medium of control for characteristics like anti-fogging abilities, peelability and ability to seal through a degree of contamination. The degree to which atmospheric modification takes place in packages is dependent upon several variables such as film permeability to O_2 and CO_2, product respiration and the influence of temperature on both of these processes.

Accordingly, there has been a lot of commercial interest to develop films with high gas transmission rates. High gas transmission films are obtained by modifying the film manufacturing process so that gases such as O_2, CO_2 and water vapour exit or enter the package in a controlled manner such that aerobic respiration needs are met and desirable CO_2 and moisture levels are maintained.

Table: Gas permeability and water transmission rate.

Film	Permeability (cm^3/m^2 d atm) for 25 m film at 25 °C			WTR (g/m^2/day/atm) at 38 °C, 90% RH
	O_2	N_2	CO_2	
Ethylene- vinyl alcohol (EVAL)	3–5	–	–	16–18
PVdC–PVC copolymer (Saran)	9–15	–	20–30	–
Low-density polythene (PE- LD)	7800	2800	42.000	18
High-density polyethylene (PEHD)	2600	650	7600	7–10
Polypropylene cast (PPcast)	3700	680	10,000	10–12
Polypropylene, oriented (OPP)	2000	400	8000	6–7
Polypropylene, oriented, PVdC coated (OPP/PVdC)	10–20	8–13	35–50	4–5
Rigid poly(vinyl chloride) PVC	150–350	60–150	450–1000	30–40
Plasticized poly(vinyl chloride) (PVC-P)	500–30.000	300–10.000	1500–46.000	15–40
Ethylene - vinyl acetate (EVAC)	12.500	4900	50,000	40–60
Polystyrene, oriented (OPS)	5000	800	18,000	100–125
Polyurethane (PUR)	800–1500	600–1200	7000–25,000	400–600
PVdC–PVC copolymer (Saran)	8–25	2–2.6	50–150	1.5–5.0
Polyamide (Nylon-6), (PA)	40	14	150–190	84–3100

For most plastic films, CO_2 permeability is 2-6 times greater than that of O_2, which provides a range of CO_2/O_2 permeability ratios. This allows the selection of films of suitable rations for some products. Thus, packaging film of the correct permeability must be chosen to realise the full benefits of MAP of fresh produce. Typically, the key to successful MAP of fresh produce is to maintain equilibrium MA (EMA) containing 2–10 % O_2/CO_2 within the package. For highly respiring produce such as mushrooms, beansprouts, leeks, peas and broccoli, traditional films like low-density polythene (PE-LD), poly(vinyl chloride) (PVC), ethylene/vinyl acetate (EVAC), oriented polypropylene (OPP) and cellulose acetate are not sufficiently permeable. Such highly respiring produce is most suitably packed in highly permeable microperforated films.

The permeabilities of two commercial ceramic-filled PELD films were measured and compared with those of a plain PE-LD film. The ceramic films have higher oxygen (O_2), carbon dioxide (CO_2) and ethylene (C_2H_4) permeabilities, higher CO_2 to O_2 permeability ratio, and higher C_2H_4 to O_2 permeability ratio. Table gives the details of film types and permeability availability. Films that have improved rates of gas transmission by virtue of their polymeric nature are usually blends of two or three different polymers, where each polymer of a blend performs a specific function such as strength, transparency and improved gas transmission to meet certain product descriptions. Furthermore, films can be laminated to achieve needed properties. Among this class are high (6–18 %) EVAC content, PE-LD, oriented polypropylene laminates, styrene butadiene block copolymer films and ultra-low-density ethylene octene copolymer films and polyolefin plastomer octene copolymer films. A lot of work has been reported on MAP using different types of films. Commercially available MAP systems for small and large quantities of produce are given in Table. Films using micro perforations can attain very high rates of gas transmission. The diameter of micro perforation generally ranges from 40 to 200 μm and by altering the size and thickness of micro perforations; gas permeability through a package can be altered to meet well defined product requirements. Based on the rates of respiration and gas transmission through the micro perforations and the base film, packages have been developed that maintain desired levels of O_2 and moisture for high respiring mushrooms. Micro perforated films have also been used to extend the storability of strawberries and nectarines, leeks, asparagus, parsnips, cherry tomatoes, sweet corn and apple. A procedure permitting the simple determination of O_2 and CO_2 permeance of micro perforated films used to pack respiring foods under real conditions has been evaluated. Macro perforated films has also been used to improve the keeping quality of strawberries and raspberries by combining high O_2 atmospheres with low oxygen MAP. However, microperforated films are relatively expensive, permit moisture and odour loss, and may allow for the ingress of microorganisms into sealed packs during wet handling situations.

A very interesting development for the packing of fresh prepared produce involves the use of high O_2 (70–100 %) MAP which has been recently shown to overcome the many disadvantages of current air packing and low O_2 MAP. High O_2 MAP has been demonstrated to inhibit enzymic discolorations, prevent anaerobic fermentation reactions and inhibit microbial growth with the result of extending prepared produce shelf-life.

Different commodities have different amounts of internal air space (potatoes 1–2 %, tomatoes 15–20 %, apples 25–30 %). A limited amount of air space leads to increase in resistance to gas diffusion. One of the primary effects of MAP is a lower rate of respiration, which reduces the rate of substrate depletion. Ethylene (C_2H_4) is a natural plant hormone and plays a central role in the initiation of ripening, and is physiologically active in trace amounts (0.1 ppm). C_2H_4 production is reduced by about half at O_2 levels of around 2.5 %. This low O_2 retards produce ripening by inhibiting both the production and action of C_2H_4.

Table: Recommended gas mixtures for MAP.

Fruits	O_2 (%)	CO_2 (%)	N_2 (%)	Vegetables	O_2 (%)	CO_2 (%)	N_2 (%)
Apple	1–2	1–3	95–98	Artichoke	2–3	2–3	94–96
Apricot	2–3	2–3	94–96	Beans, snap	2–3	5–10	87–93
Avocado	2–5	3–10	85–95	Broccoli	1–2	5–10	88–94
Banana	2–5	2–5	90–96	Brussels sprouts	1–2	5–7	91–94
Grape	2–5	1–3	92–97	Cabbage	2–3	3–6	81–95
Grapefruit	3–10	5–10	80–92	Carrot	5	3–4	91–95
Kiwifruit	1–2	3–5	93–96	Cauliflower	2–5	2–5	90–96
Lemon	5–10	0–10	80–95	Chili peppers	3	5	92
Mango	3–7	5–8	85–92	Corn, sweet	2–4	10–20	76–88
Orange	5–10	0–5	85–95	Cucumber	3–5	0	95–97
Papaya	2–5	5–8	87–93	Lettuce (leaf)	1–3	0	97–99
Peach	1–2	3–5	93–96	Mushrooms	3–21	5–15	65–92
Pear	2–3	0–1	96–98	Spinach	Air	10–20	–
Pineapple	2–5	5–10	85–93	Tomatoes	3–5	0	95–97
Strawberry	5–10	15–20	70–80	Onion	1–2	0	98–99

Also metabolic processes such as respiration and ripening rates are sensitive to temperature. Biological reactions generally increase two to three-fold for every 10 °C rise in temperature. Therefore temperature control is vitally important in order for a MAP system to work effectively. Film permeability also increases as temperature increases, with CO_2 permeability responding more than O_2 permeability. Low RH can increase transpiration damage and lead to desiccation, increased respiration, and ultimately an unmarketable product. One serious problem associated with high in-package humidity is condensation on the film that is driven by temperature fluctuations. A mathematical model was developed for estimating the changes in the atmosphere and humidity within perforated packages of fresh produce. The model was based on the mass balances of O_2, CO_2, N_2 and H_2O vapours in the package. Also a procedure to maintain desired levels of O_2 and CO_2 inside packages that are exposed to different surrounding temperatures was designed and tested.

The concentrations of O_2 and CO_2 within a package can be modeled. Useful models have been developed that would allow fresh produce processors to choose packaging materials most suited to the enclosed product. A common mathematical model involves the use of what is known as a Michaelis–Menten type respiratory model to describe the influence of temperature, O_2 (and potentially CO_2) on respiration. This approach has been used for blueberries, strawberries, raspberries and apple slices. Respiratory models are then coupled with an equation describing the temperature sensitivity of film permeability to gases (known as the Arrhenius equation) to predict package O_2 partial pressure as a function of temperature, product mass, surface area and film thickness. Using polymeric films, MA packaging systems for products with low to medium respiration rates have to some extent been successfully developed. Products such as broccoli, mushrooms, leeks, etc. exhibit very high rates of respiration such that conventional films can potentially over modify the pack atmosphere and thus result in fermentation.

Qualitative changes in the broccoli (Brassica oleraceaitalica) under MAP in perforated polymeric film had also been studied. It has been found that perforated PP film packages (2 holes, each of 0.3 mm in diameter) and having a film area of 0.1 m^2 could be used to store broccoli for 4 days under MAP with maintenance of chlorophyll and ascorbic acid. Comparative evaluation of the effect of storage temperature fluctuation on MAP of selected fruits and vegetables like mushrooms (Agaricus bisporus), broccoli (Brassica oleracea L. cv. Acadi) and mature-green tomatoes (Lycopersicon esculentum cv. Trust) had also been done by researchers.

The quality of the products stored under the temperature fluctuating regime was severely affected as indicated by extensive browning, loss of firmness, weight loss increase, the level of ethanol in the plant tissue, and infection due to physiological damage and excessive condensation, compared to products stored at constant temperature. It was clear that temperature fluctuation, even if it should occur only once, can seriously compromise the benefits of MAP and safety of the packaged produce. The effect of storage conditions and physical tissue damage on membrane peroxidation in minimally processed cucumber tissue was investigated. Tissue hardness of MAP sealed samples increased for the first 3 days for all tissues but chilled tissues started to soften on the sixth day. Cucumbers packaged in perforated 31.75 μm PELD bags were found to have less severe chilling injury than non-wrapped fruit in storage at 5 °C and 90–95 % RH. The influence of MAP on the sensory characteristics and shelf life of shiitake mushrooms (Lentinula edodes) was also studied using PE-LD, PP and macro perforated film. Sensory analysis showed that mushrooms stored under MA (active and passive) had a higher deterioration rate than those stored in PP macro perforated films, and lower sensory quality values during the entire storage time.

These results suggest that mushroom deterioration was probably due to shiitake mushrooms' sensitivity to high CO_2 concentrations. Mushrooms have been packed under micro porous and hydrophilic films and stored at 10 °C and 20 °C under high relative humidity (>92 %) to study the unique gas barrier properties of hydrophilic films (wheat gluten based material and synthetic polymer). Unique steady state atmospheres, poor in both oxygen and carbon dioxide, were observed, regardless of the temperature and the hydrophilic film used, owing to their high selectivity to gas diffusion. The effect of super-atmospheric O_2 and MAP on plant metabolism, organoleptic quality and microbial growth of minimally processed baby spinach was also studied. Packaging film O_2 transmission rate and initial levels of super-atmospheric O_2 in the packages significantly affected the changes of in-package atmospheres during storage and consequently quality of baby spinach leaves. Microbial proliferation and sensory quality aspects of sliced onions were tested at different temperatures (−2, 4 and 10 °C) and atmospheric conditions (with or without 40 % CO_2/59 % N_2/1 % O_2) by researchers. The microbial shelf lives of the tested onions in 40 % CO_2/59 % N_2/1 % O_2, or at -2, 4 and 10 °C, were 12.5, 9.5, 7, 12, 9 and 6 days, respectively, and their sensory shelf lives were 12, 8, 5, 10.5, 7 and 5 days, respectively.

Work on other vegetables has also been reported like sweet corn and snow pea pods. The shelf life of shelled peas packed in PE-LD (25 µm thickness) was 45, 17, 7 and 4 days when stored at the temperatures of -11, 5, 15 °C and room temperature respectively, considering the quality indices like total soluble solids, total water soluble sugars, protein, physiological weight loss and decay. The shelf life of shelled peas packaged in PE-HD was found as 4, 7, 17 and 45 days when stored at the temperatures of ambient, 15, 5 and -11 °C, respectively. The influence of different storage conditions on the storability of packaged shredded cabbage had also been studied. A higher variation in CO_2 and O_2 concentrations, and consequent accumulation of anaerobic metabolites had a negative influence on the sensorial properties of the cut cabbage. Nitrogen (N_2) gas packaging for fresh-cut vegetables (lettuce and cabbage) has been examined as a means of MAP for extending the shelf life of cut vegetables. Degradation of cut vegetables in terms of appearance was delayed by N_2 gas packaging. Because of this effect, the appearance of fresh-cut vegetables packaged with N_2 gas remained acceptable at temperatures below 5 °C after 5 days.

Table: Effect of packaging materials and methods on the shelf life of fruits and vegetables.

Food/Treatment	Packaging materials/methods	Shelf life
Peach, Cauliflower, Truffle	Tray: PP; Cover: PE-LD/PET(40 µm), 0-14 microperforated package, all wrapped in PE	4 days at 4 °C
Strawberry	stretch PVC	8 days at 1 °C
Minimally processed fruits (kiwi, banana and prickly pear)	1) PE/Al/PET	4-12 days at 5 °C
	2) Coex.polyolefinic high permeable film	
Sweet cherry	5 % O_2 + 10 % CO_2	80 days at 1 °C
	PE: 13-18 % O_2 + 2-4 % CO_2	40 days at 1 °C
	70 % O_2 + 0 % CO_2	20 days at 1 °C
	Air	30 days at 1 °C
Cactus pear fruits	Cryovac MY 15 Plastic box	9 days at 4 °C
Carrots, minimally processed	PP + cPP/OPP in:	
	5 % O_2/10% CO_2/85 % N_2	2 days at 4 °C
	80 % O_2/10 % CO_2/10 % N_2	7 days at 4 °C

Cabbage, shredded	OPP (30 μm)	9-10 days at 3 °C
Cabbage, shredded	Glass jar; PE (30 μm); PP (30 μm) in:	7 days at 0 and 10 °C
	Air	
	100 % N_2,	
	MAP 1: 100 % N_2,	
	MAP 2: 5 % O2/95 % N_2,	
	MAP 3:10 % O_2/90 %	
	MAP 4: 70 % O_2/30 % N_2 and 100 % O_2	
Lettuce (shredded)	OPP (30 μm)	8 days at 4 °C
Spinach Fresh-cut	PE-LD	8 days at 4 °C and 90 % HR
	OPP	6 days at 4 °C and 90 % HR
Lettuce	PE	10 days in the dark at 7 °C.
Broccoli	PE-non-perforated	13 days at 4 °C; 3 days at 20 °C
	PE-2 holes (750 μm in dia.)	23 days at 4 °C;
	PE-4 holes (8.8 mm in dia.)	5 days at 20 °C
	unwrapped broccoli	10 days at 4 °C, 2.5 days at 20 °C
Fresh-cut vegetables, fennels cut lettuce salads and shredded carrots	PA/PE (102 μm)	
	Air	less than 5 days at 4 °C
	Vacuum	13.5 at 4 °C
	MAP: 70 % N_2/30 % CO_2	5.8 at 4 °C
Cherry tomatoes ozone treatment (5–30 mg/l ozone gas for 0–20 min)	CA 1: 5 % CO_2/17 % O_2 (7 °C)	20 days at 7 °C
	CA 2: 5 % CO2/15 % O_2 (22 °C)	10 days at 22 °C
	PE-LD (50 μm); MAP	
	glass jars	
Melon, minimally fresh processed	MAP passive in:	10 days at 5 °C
	PP mikroperforated	
	BOPP	
	OPP	
	PP-makroperforated (control)	
Blueberry nectar Aseptically Filled (93–97 °C for 30 s)	PET 1: Standard	9-12 months at 4-20 °C
	PET 2: O_2 absorber	
	PET 3: 5-layers	
	PET 4: 3-layers	
Tomato	PE-LD (50 mm) MAP + O_2 absorber	8 days at 20 °C

The MAP combined with ozone and edible coating films were used for improving the effect of preservation of strawberry. The optimum gas composition of MAP test for strawberry was 2.5 % O_2/16 % CO_2. Also the integrated model approach was used to study the effect of MA conditions on the keeping quality of 'Elsanta' strawberries as limited by spoilage. It was observed that the quality of SO_2 -free 'Superior seedless' table grapes was preserved in MAP. The improvement of the overall quality of table grapes stored under MAP in combination with natural antimicrobial compounds had also been studied. The effect of MA on the quality of many fresh-cut products has been studied. Successful applications include mushroom, apples, tomato, pineapple, butter head lettuce, potato, kiwifruit, salad savoy, honeydew, mangoes and carrot.

Fresh-cut cantaloupe cubes were placed in film sealed containers in which the internal gas mixture was attained naturally (nMAP), was flushed with 4 kPa O_2/10 kPa CO_2 (fMAP), or was maintained near atmospheric levels by perforating the film (PFP). While both nMAP and fMAP maintained the sale able quality of melon cubes at 5 °C, fMAP maintained quality better than MAP. Also quality of fresh-cut tomato slices was compared during cold storage under various MAP conditions. MAP provided good quality tomato slices with a shelf life of 2 weeks or more at 5 °C. Effects of packaging materials and methods on the shelf life of some fruits and vegetables are given in Table.

Edible Coatings and Films

The application of edible coatings appears to be one of the most innovative approaches to extend the commercial shelf life of fruits and vegetables by, among other mechanisms, acting as a barrier against gas transport and showing similar effects to storage under controlled atmospheres. Indeed, over the last two decades the development and use of bio-based packaging materials to prolong the shelf-life and improve the quality of fresh products has been receiving increased attention. The reasons for such an interest are mainly related to environmental issues due to disposal of conventional synthetic food-packaging materials. However, in order such edible films and coatings to be used at a commercial level in food products they must fulfill some basic requirements: acceptable sensorial characteristics, appropriate barrier properties, good mechanical strength, reasonable microbial, biochemical and physicochemical stability, safety, low cost and simple technology for their production. The effectiveness of edible coatings for protection of fruits and vegetables also depends on controlling the wettability of the coating solutions, which affects the coating thickness. Thus, edible coating formulations must wet and spread uniformly on the vegetable's surface and, after drying, a coating that has adequate adhesion, cohesion and durability to function properly must be formed.

Table: Oxygen and carbon dioxide permeabilities of edible films.

Film	Thickness (mm)	Permeability at 0% RH (10^{-15} l/m^2 s Pa)		Permeability ratio (CO_2/O_2)
Corn-zein	0.12-0.31	0.36 30 °C	2.67 21 °C	7.5
Wheat gluten	0.23-0.42	0.20 30 °C	2.13 21 °C	9.5
Methyl cellulose Low level (MC (L))	0.04-0.07	2.17 30 °C	69.00 21 °C	31.6
Hydroxypropylcellulose Low level (HPC (L))	0.05	3.57 30 °C	143.99 21 °C	40.6
HPC/Lipids	0.15	3.44 30 °C	81.75 21 °C	23.7
Cozeen	0.09	0.89 37,8 °C	5.25 22,8 °C	5.9

Wheat gluten	0.14	0.09 37,8 °C	0.03 22,8 °C	0.3
Corn-zein	0.08	0.16 25 °C	–	–
Wheat gluten	0.15	0.08 25 °C	–	–

Edible films and coatings are generally based on biological materials such as proteins, lipids and polysaccharides. The main polysaccharides that can be included in edible coating formulations are starch and starch derivatives, cellulose derivatives, chitosan, pectin, alginate and other gums. Carboxymethyl-cellulose is a cellulose derivative that has received considerable attention with several examples of applications in many fruits and vegetables. Some polysaccharides have been used as edible coatings to improve the quality of different fresh-cut fruit.

A commercial edible coating formulation based on carboxymethylcellulose and sucrose fatty acid esters, has been applied to pears, cherries and asparagus. Pullulan an extracellular polysaccharide produced by Aureobasidium pullullans, also is capable of forming edible films but has not been largely exploited as a coating material in fruits and vegetables, presumably because of its high water solubility. One example of pullulan used as a coating hydrocolloid was for strawberries and kiwifruit. Proteins that can also be used in formulations of edible coatings for fruits and vegetables include those derived from animal sources, such as casein and whey proteins, or obtained from plant sources, like corn zein, wheat gluten and soy protein. Whey protein based coatings have been extensively used to extend the shelf life of fruits and vegetables.

Antimicrobial Packaging

The use of edible coatings as carriers of antimicrobial compounds is another potential alternative to enhance the safety of fresh-cut produce. Antimicrobial edible coatings may provide increased inhibitory effects against spoilage and pathogenic bacteria by maintaining effective concentrations of the active compounds on the food surfaces. There are several categories of antimicrobials that can be potentially incorporated into edible coatings, including organic acids (acetic, benzoic, lactic, propionic, sorbic), fatty acid esters (glyceryl monolaurate), polypeptides (lysozyme, peroxidase, lactoferrin, nisin), plant essential oils (cinnamon, oregano, lemongrass), nitrites and sulphites, among others. Although several types of antimicrobials incorporated into edible coatings have been used for extending shelf-life of fresh commodities, their use in fresh-cut fruits is yet limited. Currently, organic acids and plant essential oils are the main antimicrobial agents incorporated into edible coatings for fresh-cut fruits. Despite the good results achieved so far with the incorporation of essential oils into edible coatings, the major drawback is their strong flavour which could change the original taste of foods.

To confer antimicrobial activity, antimicrobial agents may be coated, incorporated, immobilised, or surface modified onto package materials. The classes of antimicrobials listed range from acid anhydride, alcohol, bacteriocins, chelators, enzymes, organic acids and polysaccharides. Examples of commercial antimicrobial materials in the form of concentrates (e.g. AgIONe, AgION Technologies LLC, USA) extracts (Nisaplin® (Nisin), Integrated Ingredients, USA) and films (Microgarde Rhone-Poulenc, USA) were also presented. Antimicrobial packages have had relatively few commercial successes except in Japan where Ag-substituted zeolite is the most common antimicrobial agent incorporated into plastics. Ag-ions inhibit a range of metabolic enzymes and have strong antimicrobial activity. Antimicrobial films can be classified into two types: those that contain an antimicrobial agent which migrates to the surface of the food and, those which are effective against surface growth of microorganisms without migration. Coating of films with antimicrobial agents can result in effective antimicrobial activity. Natrajan and Sheldon carried out a study to evaluate the potential use of packaging materials as delivery vehicles for carrying and transferring nisin-containing formulations onto the surfaces of fresh poultry products. The antimicrobial activity of the fabricated multilayer films was also evaluated using an agar plate diffusion method. It was reported that coating the PE film with grapefruit seed extract (GFSE) with the aid of a polyamide binder resulted in greater antimicrobial activity compared to GFSE incorporation by co-extrusion. Using the agar diffusion test, the co-extruded film with 1% w/w GFSE showed antimicrobial activity against Metaphycus flavus only, whereas a film coated with 1% GFSE showed activity against several microorganisms such as Escherichia coli, Staphylococcus aureus and Bacillus subtilis. There is a growing interest in edible coatings due to factors such as environmental concerns, the need for new storage techniques and opportunities for creating new markets for underutilised agricultural commodities with film forming properties. Edible coatings and films prepared from polysaccharides, proteins and lipids have a variety of advantages such as biodegradability, edibility, biocompatibility, aesthetic appearance and barrier properties against oxygen and physical stress.

Antimicrobial packaging can play an important role in reducing the risk of pathogen contamination, as well as extending the shelf-life of foods; it should never substitute for good quality raw materials, properly processed foods and good manufacturing practices. It should be considered as a hurdle technology that in addition with other non-thermal processes such as pulsed light, high pressure and irradiation could reduce the risk of pathogen contamination and extend the shelf-life of perishable food products. Participation and collaboration of research institutions, industry and government regulatory agencies will be a key on the success of antimicrobial packaging technologies for food applications.

Active Packaging

Active packaging is an emerging and exciting area of food technology that is developing owing to advances in packaging technology, material science, biotechnology and new consumer demands. This technology can confer many preservation benefits on a wide range of ambient-stable and chilled food products. The intention is to extend the shelf-life of foods, whilst at the same time maintaining nutritional quality and assuring microbial safety. The use of active packaging is becoming increasingly popular and many new opportunities will open up for utilising this technology in the future.

Active packaging employs a packaging material that interacts with the internal gas environment to extend the shelf-life of a food. Such new technologies continuously modify the gas environment (and may interact with the surface of the food) by removing gases from or adding gases to the headspace inside a package.

Ethylene Scavenging

A chemical reagent, incorporated into the packaging film, traps the ethylene produced by ripening fruit or vegetables. The reaction is irreversible and only small quantities of the scavenger are required to remove ethylene at the concentrations at which it is produced.

Oxygen Scavenging

The presence of oxygen in food packages accelerates the spoilage of many foods. Oxygen can cause off-flavour development, colour change, nutrient loss and microbial attack. One of the most promising applications of oxygen scavenging systems in food packages is to control mould growth.

Table: Properties of major types of Ageless oxygen scavengers.

Type	Function	a_w	Aborption speed	Capacities
Z	Decreases O_2	< 0.65	1 – 4 days	20 – 2 000 cc
S	Decreases O_2	> 0.65	0.5 – 2 days	500 – 2000 cc
FX	Decreases O_2	> 0.85	0.5 – 1 days	20 – 300 cc
E	Decreases O_2 Decreases CO_2	> 0.30	3 – 8 days	O_2: 25–200 cc CO_2: 250 – 2000 cc
G	Decreases O_2 Increases CO_2	0.3 - 0.5	1 – 4 days	NA

Humidity Control

Condensation or 'sweating' is a problem in many kinds of packaged fruit and vegetables. When one part of the package becomes cooler than another water is likely to condense in the cooler areas.

Carbon Dioxide Release

High carbon dioxide levels are desirable in some food packages because they inhibit surface growth of microorganisms. Fresh meat, poultry, fish, cheeses and strawberries are foods which can benefit from packaging in a high carbon dioxide atmosphere. However with the introduction of modified atmosphere packaging there is a need to generate varying concentrations of carbon dioxide to suit specific food requirements. Since carbon dioxide is more permeable through plastic films than is oxygen, carbon dioxide will need to be actively produced in some applications to maintain the desired atmosphere in the package.

Sulphur Dioxide

Serious loss of table grapes can occur unless precautions are taken against mould growth. It is necessary to refrigerate grapes in combination with fumigation using low levels of sulphur dioxide. Fumigation can be conducted in the fruit cool stores as well as in the cartons. Carton fumigation consists of a combination of quick release and slow release systems which emit small amounts of sulphur dioxide. When the temperature of the cartoned grapes rises due to inadequate temperature control, the slow release system fails releasing all its sulphur dioxide quickly. This can lead to illegal residues in the grapes and unsightly bleaching of the fruit.

Flexible packaging materials such as PE-LD and linear low density polyethylene (PE-LLD) when impregnated with potassium permanganate and cinnamic acid, respectively become ethylene scavengers. Fresh fruits and vegetables, such as mango, tomato, banana, and papaya exhibit more shelf-life of two to three weeks when packed in such ethylene scavenging films. The incorporation of ethylene scavengers improves the physico-mechanical properties of the packaging materials considerably.

Design of active modified atmosphere packaging for vegetables was studied by developing a mathematical model predicting gas changes and based on the following independently evaluated parameters: vegetable respiration rate, film permeability, and oxygen absorption kinetic of the absorber. A step-by-step model validation was performed on a tomatoes/low-density polyethylene pouch/commercial iron-based scavenger system at 20 °C.

Smart or Intelligent Packaging

Smart packaging uses features of high added value that enhance the functionality of the product, notably mechanical, electronic and responsive ink features, for example electronic and mechanical dispensers in which drugs are supplied and the prepared meal that automatically tells the microwave how it should be cooked. Smart packaging can be categorised into two types, those which incorporate integrated circuits (IC's) and does which do not incorporate IC's otherwise known as chipless smart packaging. Packaging that incorporate diagnostic indicators are also included in smart packaging. These can be used for such functions as monitoring vibration, acidity, tilt, shock, humidity, light, heat, time chemicals, virus or bacteria as they develop or as they are contacted.

Fresh-cut produce continues to be one of the fastest growing segments of food retailing and while conventional film packaging is suitable for lettuce and prepared salads, it cannot cope with the high respiration rates of pre-cut vegetables and fruit, leading to early product deterioration. Novel breatheable polymer films are already in commercial use for fresh-cut vegetables and fruit. Packaging films that are acrylic side-chain crystallisable polymers tailored to change phase reversibly at various temperatures from 0-68 °C are available. As the side chain components melt, gas permeation increases dramatically, and by further tailoring the package and materials of construction, it is possible to fine tune the carbon dioxide to oxygen permeation ratios for particular products. The final package is 'smart' because it automatically regulates oxygen ingress and carbon dioxide egress by transpiration according to the prevailing temperature. In this way, an optimum atmosphere is maintained around the product during storage and distribution, extending freshness and allowing shipping of higher quality products to the consumer. Intelligent packaging can change colour to let the customer know how fresh the food is and show if the food has been spoiled because of a change in temperature during storage or a leak in the packaging.

Time temperature integrators (TTI's) are devices that show an irreversible change in a physical characteristic, usually color or shape, in response to temperature history. The TTI's are expected to mimic the change of a certain quality parameter of the food product undergoing the same exposure to temperature. The TTI's presently on the market have working mechanisms based on different principles: biological, chemical and physical. For the first type, the change in biological activity, such as microorganisms, spores or enzymes is the basic working principle. The others are based on a purely chemical or physical response towards time and temperature, such as an acid-base reaction, melting, polymerization, etc.

Fresh-Check® LifeLines integrator is supplied as self-adhesive labels, which may be applied to packages of perishable products to assure consumers at point-of-purchase and at home that the product is still fresh. It is commonly referred to as having a bull's eye configuration. It is a full history indicator whose working mechanism is based on the color change of a polymer formulated from diacetylene monomers. It consists of a small circle of polymer surrounded by a printed ring for color reference. The polymer, which starts lightly colored, gradually deepens in color to reflect the cumulative exposure to temperature. The polymer changes color at a rate proportional to the rate of food quality loss: the higher the temperature, the more rapidly the polymer changes in color.

Vitsab Indicator is a full history integrator based on an enzymatic reaction. The device consists of a bubble-like dot containing two compartments: one for the enzyme solution, lipase plus a pH indicating dye compound and the other for the substrate, consisting primarily of triglycerides. The dot is activated at the beginning of the monitoring period by application of pressure on the plastic bubble, which breaks the seal between compartments. The ingredients are mixed and as the reaction proceed a pH change results in a color change. The dot, initially green in color, becomes progressively yellow as product approaches the end of shelf-life. The reaction is irreversible and will proceed faster as temperature is increased and slower as temperature is reduced. The integrator, in the format of adhesive labels, is available in two basic configurations: single or triple-dot. Single-dot tags are used for transit temperature monitoring of cartons and pallets of product and for consumer packages as well.

ripeSense is the world's first intelligent ripeness indicator label. ripeSense evolved from the simple idea of making a fruit label that is capable of more than just branding product and this has lead to the next revolution in fresh produce marketing.

The 3M Monitor Mark uses a coloured ester and phthalate mix with the desired melting point that is coloured with a blue dye. Above its melting point it diffuses along a wick, and the progress along this wick gives an indication of how long the indicator has been liquid. A polyester strip keeps the liquid away from the wick, until it is pulled by the operator to start the device.

It is believed that tomorrows food packages will certainly include radio frequency identification (RFID) tags. RFID tags are an advanced form of data information carrier that can identify and trace a product.

Biodegradable Packaging

Currently, there are several types of bio-based polymers on the market: same coming from petrochemical monomer, like certain types of polyester, polyester amides and polyvinyl alcohol, produced by different manufacturer, used principally as films or moulding. Four other bio-based polymers are starch materials, cellulose materials, polylactic acid (Polyester, PLA), polyhydroxy acid (polyester, PHA). Until now, the PHA polymer is a very expensive polymer because it is commercially available in very limited quantities. PLA is becoming a growing alternative as a green food packaging materials because it was found that in many situations it performs better than synthetic ones, like oriented polystyrene (OPS) and PET materials.

There is an increasing demand for identifying biodegradable packaging materials and finding innovative methods to make plastic degradable. Biodegradation is the process by which carbon-containing chemical compounds are decomposed in the presence of enzymes secreted by living organisms. The use of bioplastic is to replicate the life cycle of biomass by conserving the fossil fuels, carbon dioxide and water production. There are three requirements for the fast degradation process viz. temperature, humidity and type of microbes.

In the short term, biobased materials will most likely be applied to foods requiring short-term chill storage, such as fruits and vegetables, since biobased materials present opportunities for producing films with variable CO_2/O_2 selectivity and moisture permeability. However, to succeed, biobased packaging of foods must be in compliance with the quality and safety requirements of the food product and meet legal standards. Additionally, the biobased materials should preferably preserve the quality of the product better and longer to justify any extra material cost.

Table: Properties of some biodegradable plastics.

Material	Film preparation	Moisture barrier	Oxygen barrier	Mecanical properties
Starch/Polyvinyl alcohol (PVAL)	Extrusion	–	+	+
Polyhydroxybutyrate/valerate (PHB/V)	Extrusion	+	+	+/–
Polylactic acid (PLA)	Extrusion	+/–	–	+

Application of Nanocomposites

Nanocomposite materials are composed of nanoscale structure that enhances the macroscopic properties of food weight) of nanoclays reinforcement. However, most work done on polymer/clay nanocomposites has focused mainly on synthetic polymers.

The functional properties of biopolymer-based edible films or coatings have been shown to act as barrier to solute and gas and enhance food quality and shelf life. However, these films do not display good mechanical and water vapor barrier properties due to their hydrophilic characteristics. To overcome these issues, a new approach has been developed, which use hybrid materials consisting of polymers and layered silicates. Layered silicates, such as montmorillonite (MMT) clay mineral, result from the stacked arrangement of negatively charged silicate layers and contain a platelet thickness of about 1 nm with a high aspect ratio (ratio of length to thickness). The layered silicate filled polymer composites exhibit extraordinary enhancement of mechanical, thermal and physicochemical properties at a low level of filler concentration in comparison to pure polymer and conventional microcomposites. In particular, these nanocomposites have excellent barrier properties because the presence of clay layers delays the diffusing molecule pathway due to tortuosity.

Some of the works done with biopolymer-based nanocomposites were based on starch or polysaccharides, such as wheat and maize starch, thermoplastic starch, and chitosan. A few studies on protein-based nanocomposites including soy protein whey protein and wheat gluten. Most of the biopolymer-based nanocomposites have shown appreciable improvements in mechanical and barrier properties compared to the counterpart biopolymer films. Whey protein has received much attention for its potential use as an edible film and coating because it has been shown to make transparent films and coatings that can act as excellent oxygen barriers and provide certain mechanical properties. Unlike chitosan film, whey protein films have not shown any antimicrobial activity; therefore, incorporation of antimicrobial agents, such as sorbic acid, p-aminobenzoic acid, and lysozyme, is needed to impart this property. Recently, researcher found that chitosanbased nanocomposite films blended with some organically modified MMT, such as Cloisite 30B, exhibited antimicrobial activity against Gram-positive bacteria. They postulated that the antimicrobial action came from the quaternary ammonium salt of the organically modified nano-clay.

Packaging of Dairy Products

The appropriate packaging of milk is of utmost importance not only to preserve its nutritive value and saving of wastage, but also to improve the marketability to achieve better returns. The challenge to the packaging industry is to deliver the nutritious milk to the consumer in most economical, hygienic, safe and environmentally friendly package. Because of the shelf-life requirements of UHT milks, multiple laminates, for example a triple layer of high density polyethylene are used. An intermediate light barrier may be incorporated, according to the manufacturer's preference. Oxygen penetration during storage is not usually a problem. Whatever the source of bottles, aseptic techniques are used in filling and sealing to prevent contamination, and milk can be expected to have a shelf-life of 6 months at ambient temperatures.

Cream contains a high percentage of fat, so it is very susceptible to spoilage. So packaging material used should be selected in such a way that it possesses good grease resistance, and barrier properties against oxygen and moisture. Cultured milk products have various textures and viscosities, i.e. when liquid they are considered as beverages (Butter milk) and when semisolid they are spoonable products (Yoghurt). The material used for packaging must be compatible with the special physical chemical and bacteriological properties of fermented milk.

Traditionally, Indian dairy products have been manufactured by individual sweet makers- "halwais" and small entrepreneurs. Very little attention is paid to sanitary handling and packaging of these products. Indigenous products have been packed in leaves, paper cartons or paper-board boxes. These materials do not provide sufficient protection to the product from atmospheric contamination and manual handling. Consequently, the sweets soon lose their typical body and texture, absorb foreign odours, lose their aroma characteristics and show mold growth. Moreover, the products are stored in open metal trays.

Packaging of Fluid Milk

There are two main types of packaging systems for fluid milk one is traditional bottling system in which container is to be returned and other is one way delivery in which container is disposable and does not travel back to the dairy. In the non-returnable distribution system there are several alternative systems where different packaging materials, shapes, sizes, forms and machines are employed for packs.

For selection of a suitable package material, the knowledge of important characteristics of milk/ milk products is essential.

Characteristics of Pasteurized Milk

- Milk has a tendency to absorb the flavours from its environment.
- Risk of Contamination is more in liquid milk.
- Adulteration can be done easily when not packed properly.
- It is difficult to handle milk in bulk quantities.
- Milk is prone for oxidation when exposed to sun light.

Criteria for Selection of Milk Packaging Material for Fluid Milk

Milk, an extremely perishable and sensitive product, need exact packaging material in order to preserve its initial quality for some span of time. The necessary characteristics of packages for pasteurized milk are:

- It should be free from off-flavours.
- It should not impart any taste or flavour to the product.
- It should act as barrier to bacterial contamination.
- It should be resistant to UV light (max transmission: 8% at 500 nm and 2% at 400nm).
- It should have no physiological effects on the products.
- It should possess good mechanical properties (sealing, tensile, structural strength etc.).
- It should be tamper proof.
- It should possess good oxygen barrier properties.
- It should be economical.
- It should fit in to processing- in-Line.

Materials Used

- Glass.
- Plastics: (1) LDPE is widely used (2) LLDPE: 25% thinner film used and LDPE and LLDPE in 5:1 to 4:1 ratio.
- Others: Coated paper board, wax coated paper board ESL (Extended Shelf life Pouch).

Glass Bottles

Packaging of milk in glass bottles is the oldest system. Clear glass bottles of 500ml capacity conforming to IS: 1392 – 1967 are used. The glass bottles offer certain advantages like transparency, rigidity, hygienic and non-toxic nature and compatibility. This system involves collection and transportation of empty bottles to processing plant, washing and sterilization of dairy bottles. This packaging system requires large storage space requirement for both empty as well as filled bottles. These factors increase the fixed and the variable costs. Heavy weight, fragility and return ability of bottles has inconvenience both to the distributor as to the well as consumer which made this packaging system undesirable and is not in use at present.

Returnable Plastic Bottles

Plastic bottles reduce the weight and the chances of breakage are rare but most other characteristics of packaging milk in glass bottles remain same.

Non-Returnable Plastic Bottles

This system reduces transport charges through light weight and one way of delivery of bottles in a convenient way in comparison to any other system. Ex: HDPE with PP lid.

Plastic Films

Plastic pouches are generally made of low-density polyethylene (LDPE) film. Coextruded LDPE-LL-DPE film is also used because of its advantage of eliminating pin-hole problems. The films are of 45-75μ thick. The pouches are formed and filled on form-fillseal (FFS) machines in capacities of 200, 500ml and 1000 ml. The film should confirm to IS: 11805 – 1999. The plastic pouches are clean, hygienic and safe for this application and since these are not reused, the cleaning operation is eliminated and energy loss is avoided. Moreover, these pouches are easily recyclable for other purpose.

Aseptic Packaging of Milk

Aseptic or long-life milk was originally introduced in Sweden in the early 1960's originally called the "Tetra-pack" system. It utilizes a laminate pre sterilizer and a filling environment heater. Aluminium foil is an integral part of the flexible laminate in order to provide a barrier against light and gas. In UHT processing, Milk is preheated to 73– 850C then rapidly raised to 135 °C for fraction of second and then suddenly cooled by flashing into a vacuum chamber. It must be packed under completely sterile conditions. No refrigeration is necessary for at least 3-6 months. If kept under refrigeration a self-life of up to 1 year is possible. In the distribution system, the pouches are placed in reusable multi-trip plastic crates. The crates are made of HDPE or PP and packs are stackable.

Paper Board Cartons

This kind of packaging materials are commonly used in aseptic filling systems for milk, cream, fruit juices, soups etc. The filling systems could be either of the following two types: those in which the carton is formed within the filler from a continuous reel of material; and those in which the cartons are supplied as preformed blanks, folded flat, which are assembled into cartons in the filler. The packaging material is mainly composed of printed- Paper coated with aluminium foil and several plastic layers (Polyethylene - paper board – polyethylene - aluminium foil - polyethylene). The inner material side of the finished package is coated with a special layer facilitating the sealing process. Each layer has a specific function:

- The outer polyethylene layer protects the ink and enables the sealing process of the package flaps.
- The paperboard serves as a carrier of the decor and printing gives the package required mechanical strength.
- The laminated polyethylene binds the aluminium to the paper.
- The aluminium foil acts as a gas and light barrier.
- The inner polyethylene layer provides liquid barrier and sealing.

The same type of packaging material can be used for packaging of flavoured milk also depending on treatment and shelf-life.

Packaging of Cream

Cream is the concentrated form of milk fat. Cream contains a high % of milk fat and is very susceptible to spoilage. Moisture loss from cream can occur if not properly packed. It is prone to oxidative and lipolytic rancidity. It can absorb flavours.

Characteristics of Packaging Material Required

- Prevent light passage.
- Prevent water loss.
- Prevent oxygen transmission.
- Shall offer resistance for microbial contamination.

In early 20th century waxed paperboard cartons were used as containers for cream. Now a days creams is packed in similar packages used for milk i.e. Newer cream packaging concepts include thermoformed packs made from linear polyethylene, polystyrene or polypropylene. These may be closed with a peelable lid or snap-on cover. Tin plate containers have also been used for larger sizes. Whipped cream and synthetic formulations are sold in aerosol cans and polyethylene tubes.

Imitation cream made from soybeans and vegetable oils is often marketed in wax coated paper board cartons. Ultra pasteurization has been applied to heavy and light creams. The product then goes for packing. A strong seal is necessary for product protection. PE extruded or wax coated paperboard tubs are used to pack single portion cream. Sterilized/UHT cream is packed in similar lines to that of UHT milk.

Packaging of Fermented Dairy Products

The material used for packaging of fermented dairy products must be compatible with the special physical chemical and bacteriological properties of fermented milk. The packaging materials used are glass, polyethylene, complex card, LD/LLDP, HDPE cups, Co extrusion plastic complexes, polystyrene- EVOH (polyvinyl alcohol and ethylene co polymer, Polystyrene –PETG (polyethylene glycol terephthalate).

1. Yoghurt: Yoghurt has become very popular in many nations. In the United States yoghurt is packed in coated paper board containers. The product is automatically filled in polystyrene tubs and covered with 0.005mm aluminium foil. The inherent brittleness of polystyrene may become a problem. The terpene containing aroma of fruit yoghurt has an effect on polystyrene. Fruit acids may also cause pitting of unlacqurered aluminium lids. A polystyrene tub is an excellent package for yoghurt. It is economical, practical and widely used. Yoghurt is also being sold in expanded polystyrene foam tubs in trays. Each tray holds 20 tubs of yoghurt. Trays are stackable and serve not only as display holders and light weight shippers but also hold during fermentation of yoghurt. This saves extra handling and repacking cost.

2. Fermented milks (Buttermilk, sour cream, lassi): LDPE sachets, polystyrene cups, polypropylene cups are used. Wax coated paper cups are used. Recently buttermilk, lassi and sour cream are packed in aseptic tetra packs. HIPS for stored products must never be employed as free fatty acids may lead to cracking of the material. Lassi which is a sweetened butter milk product is packed in polystyrene cups with coated aluminium foil lids. The packaging materials such as earthen wares give firmness to product. Others are glass jars, PS/PP cups, PVC lined HDPE. Tetrapak/Brick are recommended for Lassi.

3. Dahi: The traditional pack so far was the earthenware pot with a loose cover of glassine or greaseproof paper. The earthenware pots are very heavy, easily breakable and because of oozing of water from its body, the product inside develops shrinkage cracks. Recently, injection moulded polystyrene and Polypropylene/HIPS cups have been introduced with aluminium foil based peelable lids. These cups are available in capacities of 200 grams and 400 grams and provide a shelf-life of about 10 days under refrigeration. The plastic cups are light in weight, easy to handle and are hygienic. Some private dairies are also packing dahi in LDPE pouches of 200ml capacity.

4. Shrikhand: Shirkhand is packed in polystyrene/PP Cups. HIPS are more common for 100 gm to 1 kg size which are sealed/capped/lined with Al-PE foil. Small manufacturers use lined (Glassine) paper board boxes. HDPE containers with lid of LDPE made by injection moulding are also in use. PP or PE bags, Glassine paper lined containers are also used.

Packaging of Ice Cream

The chief requirements of packages for ice cream are protection against contamination, attractiveness, ease of opening and reclosure and ease of disposal, protection against moisture loss and temperature fluctuations is desirable. Bulk ice cream is packed and hardened at a low temperature. Shaped bars are hardened prior to packaging.

Package Forms for Ice Cream

Most bulk ice cream is packaged in a liner less bleached sulphate board carton, coated with wax or polyethylene wax blends for protection from moisture and oxygen. Once the carton is opened it is difficult to reclose and the paper board tends to wrap. Although economic considerations favour the simple rectangular paper board carton, improved packaging often leads to higher sales and a marginal price differential may be offset by higher throughput. Other forms used for packaging of ice cream are Aluminium foil cartons, cylindrical containers, Plastic containers etc.

Packaging of Traditional Dairy Products

1. Khoa: Under existing trade practice, producers and traders do not employ any packaging for khoa. By employing proper packaging the shelf life of khoa can be enhanced. Hot packaging of khoa in pre-sterilized cans can improve the shelf life up to 14 days at room temperature and 75 days under refrigerated temperatures. Three times increase in shelf life was claimed by packing khoa in rigid polypropylene containers with lid and khoa packed in pre-sterilized laminate pouches (paper, Polyethylene and aluminium foil). Vacuum packaging and packing in Cryovac Shrink wrap pouches will prevent growth of aerobic microorganisms. However, these two packaging methods do not offer protection against bacteria growing inside khoa. Bulk packaging of khoa is done in PP buckets.

2. Peda: Peda is generally packed in paper board containers lined with parchment paper or grease proof paper. However, plastic trays, tubs can be used to pack peda to enhance its marketability. Paper board lined with PE or PP liner or glassine can be used. Peda packed in multilayer transparent laminate pouches under modified atmospheric packing of Nitrogen and CO_2 has a shelf life of 15 days at room temperature and 30 days at 20 °C. Peda samples packed with oxygen scavenger exhibited a shelf life of 2 month at 37 °C and 6 months at 20 °C.

3. Paneer: Vegetable parchment paper and PE bags are generally used. PE gives greater keeping quality (7 days at 5 °C) than that given by vegetable parchment Paper. The Cryovac system using shrink film is being successfully used. Retortable tins are also used. Long life can be given by Metallized polyester or Nylon – PET/METPET/PE or Aluminium foil or Nylon or LDPE/LLD.

Paneer is packed in tins along with the brine. These tins are sterilized and it may be having a slight cooked flavour and maillard browning which will increase with storage period. Paneer is also vacuum packed in laminated pouches to have an extended shelf life. Paneer is also packed in EVA/EVA/PVDC/EVA film under vacuum which may have a shelf life of 3 months under refrigeration.

4. Rasogolla: Tin cans with resistant lacquer + SO_2 (100 ppm maximum) are generally used which gives highest shelf life to the product. Earthen pots which may be lined with leaves are also used in some areas of the country. HIPS or PP cups are also used along with heat sealable caps. Retailers use HIPS/polypropylene cups with press on lids.

Packaging of Butter and Ghee

Butter and ghee contains a high percentage of fat, so they are very susceptible to spoilage. So packaging material used should be selected in such a way that it possesses good grease resistance, and barrier properties against oxygen and moisture. Cheese is fermented dairy product, having low pH and comparative higher moisture content while milk powder is dried product, containing very low moisture.

Butter

It consists primarily of about 80% milk fat, 16% moisture and in table butter up to 3% common salt. Because of high moisture content butter is susceptible to mold growth and lypolytic rancidity.

Characteristics of Butter

- Due to high moisture content butter unlike solid fats is susceptible to mold growth.
- Flavour and odour are easily absorbed by butter from its environment.
- Deterioration of the butter may take place due to rancidity.
- Butter has tendency to lose Moisture.

Requirement of Packaging

- Non-toxic, not harmful to consumer's health.
- It should be grease/moisture proof.

- Shall be barrier for Oxygen.
- Low metallic content as metals favor oxidation of fat.
- Shall not transmit light.

Packaging Material

Butter is packed in bulk as well as in retail packages. For bulk packaging, there is no standard method, and generally polyethylene bags/parchment papers along with corrugated boxes are used. Al-foil 0.09 mm thick, surface treated with lacquer for protecting against corrosion: Alfoil/parchment or glassine paper (40-42 gsm): PVC or cardboard with a parchment insert can be used.

Indian Standard 2034 - 1961 gives specifications for tin cans of 200 g and 400 g capacity that are to be used for package of butter. They specify tinplate thickness of 0.24 mm and 0.27 mm respectively and minimum tin coating of 17 g/m^2, besides many other requirements. Though tinplate containers are the best for product protection, owing to their high cost very little quantity of butter is packed in the tin containers. Flexible packaging materials like vegetable parchment paper or grease proof paper, aluminum foil, and paper board cartons which together give similar protection to the product are more commonly used.

Indian Standard 7161 - 1973 gives specifications for vegetable parchment paper or Grease Proof paper/Aluminum foil laminate for wrapping butter. As vegetable parchment paper has good wet strength, generally paper of 45 gsm and above and aluminum foil above 0.009 mm thickness are used. As butter is highly susceptible to foreign odour, care must be exercised while choosing adhesive and printing inks used in the manufacture of the laminates.

Cartons protect butter while handling after packaging in primary Wrapper, in the distribution system. Since butter is stored in the refrigerator, cartons may be waxed with about 10 gsm wax on each side though it is not mandatory. Paperboard can be extrusion coated with PP. Injection moulded pots and tubs of PP can also be used for packaging of the butter. High-impact polystyrene or HIPS is also used in multilayer sheet extrusion with a variety of other polymers, like PE, PP, PET, PVDC and EVOH.

Large packs of 10, 20 and 50 kg butter are packed formerly in wooden barrels/boxes or parchment paper lined corrugated boxes. For better handling, easier storage, more efficient use of storage space and economy Fibre board boxes are introduced which are lined with parchment paper.

Latest packaging material that are being used are Shallow, 1-2 mm thick Al-foil trays with heat sealable PVDC-cellophane or other suitable barrier material. Aluminum PVDC/PS cups can also be used for butter. Butter chiplets are packed in lacquered Aluminum foil.

The standards for vegetable parchment paper used for the butter packing are:

- Grammage: 41-45.
- Bursting strength: 1.8 ± 0.2 kg/cm^2:
    - Wet strength: 0.8 ± 0.2 kg/cm^2.
    - Grease resistance: Should pass the turpentine oil test.

- Acidity: 0.02% as H_2SO_4.
- pH of H_2O extract: Not less than 5.0.
- Brightness: 79.

Ghee

It is usually 100 per cent fat with little moisture (< 0.3 %), obtained by boiling butter at 110 °C till all water is evaporated with a grainy texture and a characteristic flavour. The product needs to be protected from chemical spoilage and rancidity caused by oxygen, light, heat, moisture and metal ions.

Characteristics of Ghee

- Easy to absorb flavour from its environment.
- Easily prone for oxidation.
- Prone for adulteration.

Requirements of Packaging Material

- Good fat resistance.
- Barrier properties against oxygen and moisture.
- It shall be temper proof.

Packaging Material

A major portion of ghee is packed in lacquered tinplate containers of capacities ranging from 200 ml to 15 litres/kilograms. Since the product is very sensitive to oxygen, the tinplate containers are filled to the brim without any air gap. Ghee packed in tinplate containers is fairly stable and has a shelf-life of about one year.

Alternate packages, which are plastic based, are now gradually replacing tins. For shorter shelf life, 200 ml, 500 ml and 1 litre capacity pouches made of polyethylene film, multi–layer coextruded films of LDPE/HDPE are used, which are economical. Aluminium foil laminate standby pouches are also commonly used for packaging ghee.

For long – term storage, stainless steel containers or tinplate cans are desirable. Ghee is also marketed in lined cartons with flexible laminated plastics as inner liner materials and in tetrapaks. In both these packs long shelf-life is achieved. Laminated pouches of metallized polyester based films are also used. Generally, plastic pouches are filled on automatic FFS machines. However, if the sealing surface is contaminated with the product, sealing of the pouch becomes difficult.

Recently, it is packed in certain laminates and Bag - in - Box containers which comprises of a pre-sealed bag made of polyethylene and polyamide laminates fitted with a spout and cap housed in a CFB/Duplex board box. The bag consists of two plies which is sealed together on all four sides and the spout and cap assembly is heat sealed onto it. The bag is vacuum filled and inserted manually into the box. Seven layer Nylon containing self-standing pouch with closure is also used. A laminate of HDPE/LDPE is used for packing ghee.

Another form is consisting of a multi-ply collapsible bag with a tap which can be housed in a rigid outer container. The container can be a box, a crate or a drum whose capacity varies between 3 and 200 liters. The bags and boxes are in collapsible form.

Cheese

Cheese is enzymatically coagulated and fermented dairy product, having pH between 5.0 to 5.5 and contains around 35 to 80% moisture depending on the type of cheese. Hence, it is prone to mold growth on the surface.

Packaging Material

Any material to be used for packaging natural cheese must give general protection, prevent moisture loss, improve appearance, protect against micro-organisms and prevent oxygen transmission.

Cheese is essentially a product with high fat and moisture content. Therefore, package used for cheese should prevent oxidation and mould growth. It should also have fat and grease resistance and be able to protect against micro-organisms. Oxygen is eliminated by packing cheese in hermetically sealed containers in vacuum or inert gas atmosphere. Processed cheese is usually packed in aluminum foil in cubes with different shapes. Tinplate cans are used for 200 g and above quantity. PVDC coated plastic films are suitable for cheese packaging as they provide good oxygen and moisture barrier properties.

Cream cheese is packed in foil lined card board boxes of heat stable plastic packs. Saran is used as wrapping material for Neufchatel cheese. Air evacuation and gas flushing is used for cottage cheese, green cheese is packed by waxing and paraffining or alternatively vacuum packed in polyethylene. Ripened cheese is packed in laminated cellophane film. Cheese consumer packs are generally consists of Lacquered metal cans of laminated consisting of Nylon/PVDC Copolymer or polyster/PVDC copolymer or Nylon/Polyethylene copolymer.

Processed cheese is packaged hot metallic containers. Wax coated cellophane, aluminum foil, polypropylene, PE, PVDE material is also used for packing processed cheese.

Milk Powder

The shelf life of a dehydrated product is influenced to a large extent by the packaging, which must conform to certain special criteria. These are:

- Protection of the dehydrated product against moisture, light, air, dust microflora, foreign odour, rodents etc.
- Strength and stability to maintain original container properties through storage, handling and marketing.
- Size, shape, appearance to promote marketability of the product.

- Composition of the container must be approved for use in contact with foods.
- Lower cost.

Selection of Packaging Materials

Dehydrated dairy products are generally hygroscopic in nature and even slightest increase of the moisture content will decrease the shelf life considerably. Hence, the packaging material should be impervious towards water vapour and such property is to be considered important for packing such products. When retention of low moisture content is the limiting factor of the shelf life of the product, the tests required to be made for determining the shelf life are:

- Determination of the normal moisture content of the product.
- The moisture content of the product at which the product becomes unacceptable to the consumer.
- The R.H. and temperature of the surrounding atmosphere at which the equilibrium moisture content is maintained in the product. Initial level to the level of unacceptability is calculated. From this data along with requirement for gas permeability etc. and utilizing the published data for different packaging material, the suitable packing material is selected.

Properties of Dried Milk Products Relevant to Packaging

- Hygroscopicity is the principal requirement which is important in small packs size where ratio of pack surface area to product is high. In tropical countries, because of the high humidity prevailing, this is the important factor.
- Cohesivity or the Cohesion of product is the serious stickiness problem particularly in high speed packaging lines. In case of powder, observations made on cohesivity are:
    - Cohesion increases with decreasing particle size.
    - It is independent of fat content in the range of 20-40% Fat.
    - A small amount of surface fat is sufficient to give cohesion to WMP.
    - Increasing the moisture from 2 to 4%, cohesion first decreases and beyond 4% cohesion increases very fast.
- O_2 Sensitivity: Preheating releases the -SH groups which protect against oxidization. However still there are chances of oxidation. Therefore for Fat containing products prevent the gas diffusion by employing either Vacuum or N_2 flushing.
- Light sensitivity: Powder exposed to light for long period bleaches the surface and thereafter it accelerate Fat oxidization.
- Heat sensitivity: Affect Flavour and Solubility. Ideal storage conditions are 18-19 °C temperature and dry atmosphere.
- Bulk Density is very important because Packaging size is affected. The Free flowing properties improves with increased Bulk Density.

- Odour pick up: Product containing Fat is prone to pick up odour. High Fat powders pick up odour from packaging materials, surrounding atmosphere, stores, water and houses. The off flavour problem is encountered from kraft paper, PE film, CFB, fiber board cases and even from rubber sealing compositions.
- Static electricity: The problem is difficult to resolve. The plastic packaging materials are also important where antistatic agents are added e.g. Glycol alkyl esters (prevent electrical charge accumulated on the film surface).
- Bacterial aspects: Powders of normal moisture and RH do not give rise to bacterial problem.

Requirements of Package for Dried Milk

- Adequate mechanical strength to withstand damage during packaging, handling, transportation and storage is necessary.
- Resistance to climatic hazards: The material should be resistant to damage by exposure to high/low temperature and humid atmosphere. This is more important in tropical countries.
- Convenient closure: Sealing is of supreme importance. Simple, effective re-closure is also desirable.
- Bulk packages should be light in weight, easy to handle and stack during transport and storage. Empty packages should occupy minimum storage space before use.
- Very low Water vapour and gas permeability: Dried milk absorbs moisture very easily. Powder with > 5 % moisture gets deteriorated during storage. The stale and gluey flavours result from Maillard reaction. It results in losses of solubility, colour change; lumping and free flowing properties are affected. For long storage, vacuum and N_2 flushing is essential especially in hot climate. An impervious container is needed which should also be odour-proof.
- Impermeability to light: To avoid surface bleaching and fat oxidation.
- Inertness, durable, safe, utilizing minimum space, identification of product and directions for use, easy availability at reasonable cost are the other requirements.

Packaging Materials Available

- Glass: Bottles, barrels, jars, etc.
- Metal: Cans, barrels, drums, bins, etc.
- Wood: Drums, bins, cask, barrels, etc.
- Paper and paper derivatives.
- Metal foils.

- Thermoplastics and their derivatives.
- Composite films:
    - Co-polymer film,
    - Coated film,
    - Co-extruded plastic film,
    - Laminates.

Flexibles for Powder

- Cartons lined with Al-foil-PE:
    - Bag-in-box: Coated Al-PE bag or plastic coated paper inside cartons.
    - Modified Atmosphere Packaging -80% N_2 and 20% CO_2 where O_2 content is reduced to 3.0%.
- Long storage: Lacquer/print/print pre-lacquer, Al-foil (10 gsm)/Adhesive coated paper (40 gsm)/PE (30 gsm).
- Shorter Keeping quality: Al-foil may be omitted. Paper thickness increased to ~ 70 gsm and coated with PVDC (30 gsm).
- Bulk Packaging: Sacks made of craft paper laminates, parchment, polyethylene, Al, Cellophane, bitumen, wax or paraffin. Even cast films now are used recently. Most common types:
    - WMP: Laminates of paper, PE, Foil, Metallized, BOPP/PE or PET.
    - SMP: HDPE or LDPE or Laminates of HD, LD, LLD, Nylon, Saran (coating), EVA. PE coating or wrinkling- Bag-in-box paper board/lined with paper, Al or Metallized/PE.
- Powder packaging materials:
    - Number of Kraft paper layers for strength required is 3-6 for 25 kg wt.
    - Outer Kraft paper bag strength 70 gsm, Second Kraft paper bag with paper alone. PE lined (95 gsm).
    - WMP: 4-6 ply kraft paper, 3 mm PE.
    - SMP: 2-4 ply kraft paper 2 mm PE.
    - Separate PE liner/inner bag – 0.04 mm for 25 kg and 0.05 mm for 50 kg.
    - PE varies in thickness from 0.02 to 0.08 mm as inside layer.

Retail Packages

- Thick laminated paper (45 gsm) – Al-foil (9 μ) – PE (25 μ) having Bursting strength 179 KPa (1.83 kg.f/cm^2) and very low WVTR and GTR.

- PET (12.5 μ)/Al-foil (9 μ)/PE (64 μ), Bursting strength 290 KPa (2.95 kg.f/cm²).
    - Form – Fill – Seal: 17 μ PET/9 μ PE – 9 μ foil – 70 μ PE.
    - Metallized film is also used. Lined cardboard – Adhesive/Coating of PVDC or RHC – Metallized PET/Al-foil/PE.
    - Metal/Plastic laminate – Retain O_2 content of 0.5 % even after 13 numbers. Bulk: Heavy gauge PE used mainly for sea voyage (passing through tropics).
- PE is used to much lesser extent as a separate bag within or as a liner for (a) Card board cartons, (b) Calico bags, (c) Jute bags with paper, (d) Liner between jute and PE.
- Alternative material to PE:
    - Multi walled paper sack may incorporate one layer of waxed paper which is more satisfactory than paper alone but is inadequate for long storage.
    - Multi wall sack may include a layer of bituminized paper, often the outer layer with 4-5 inner kraft layers. This is a good packaging material used even for export but is slightly inferior to PE.
- Perfect closure is required:
    - Metal is completely impervious but closure is a weak point.
    - Sack is sealed by sewing threads which makes holes and therefore it is covered by H_2O proof tape.
    - Gas packaging by mixture of N_2 + H_2 + Palladium as a catalyst and if kept impervious can have up to 10 years of storage life.

Whole Milk Powder and Baby Food

Generally packed in Lacquered tins under N_2 gas packaging. Whole milk powder is packed in 15 kg tins. For retail milk powder s packed in laminate bags made of PET/Al/PE.

Skim Milk Powder

Skim milk in bulk quantities of 25 kg is packed in kraft paper bags with inside polyethylene bag. Skim milk powder for retail is packed in HDPE bottles and HDPE bags.

Malted Milk Food

The malted food beverage industry is popularly known as the health beverage sector. Historically, malted beverage has a strong association with milk. The Indian health beverages market is divided into white and brown health drinks. White beverages contribute about 65% of the market.

Malted milk foods are highly sensitive to moisture and are prone to oxidative changes in the presence of light, heat and oxygen. Aroma retention of the product and prevention of moisture and oxygen ingress are important and therefore are very critical in protecting the product, and in selection of the right packaging material.

Malted milk foods are packed in quantities of 200 grams to 1 kilogram in a variety of packages. The types of packages used conventionally are glass jars, tinplate containers, which are now slowly being replaced by plastic containers and flexible laminated pouches. Though glass containers are hygienic and safe and offer the advantage of a long shelf-life, it has the disadvantage of being heavy, fragile and costly. Tinplate containers though provide good protection from gases and moisture, are likely to rust at body welding or at top and bottom seams. Tinplate containers are also expensive. The plastic containers used are blow moulded HDPE or HM– HDPE or stretch blow moulded PET containers/jars, for brown and white malted milk food products for capacities ranging from 200 grams to 2.5 kilograms. The plastic containers are light weight, sturdy, unbreakable and hygienic, and have a good shelf appeal.

Besides the plastic containers, the trend is also to use flexible pouches, which may be with or without paperboard cartons. This type of pack has an advantage of low cost compared to plastic bottles. Moreover, the storage space requirement is low and the filling operation is comparatively faster. Some of the typical structures of flexible materials used are:

- 50 and 100 grams:
    - 12μ PET/12μ metallised PET/38μ LDPE.
- 500 grams:
    - 12μ PET/20μ metallised BOPP/50μ LDPE.
    - 12μ PET/12μ metallised PET/50μ LDPE.
- 1000 grams:
    - 12μ PET/12μ metallised PET/65μ LDPE.

Packaging of Beverages

Carbonated Beverages in PET Bottles.

The beverage industry is one among the front-liners where massive investments are being made for expansion and technological up-gradation. The packaging of beverages both carbonated and non-carbonated, is a complex technological branch in the Food Processing/Packaging industry. The traditional returnable glass bottle has given way to newer plastic containers as well as cartons.

The current trend is to improve the conventional containers, extend their share in the large market, extend the shelf-life of the products, provide greater consumer convenience and ultimately to produce economic packages.

Table: Production of Soft Drinks.

Year	Bottles(in million)
1997-1998	4920
1998-1999	5670
1999-2000	6230
2000-2001	6450
2001-2002	6600

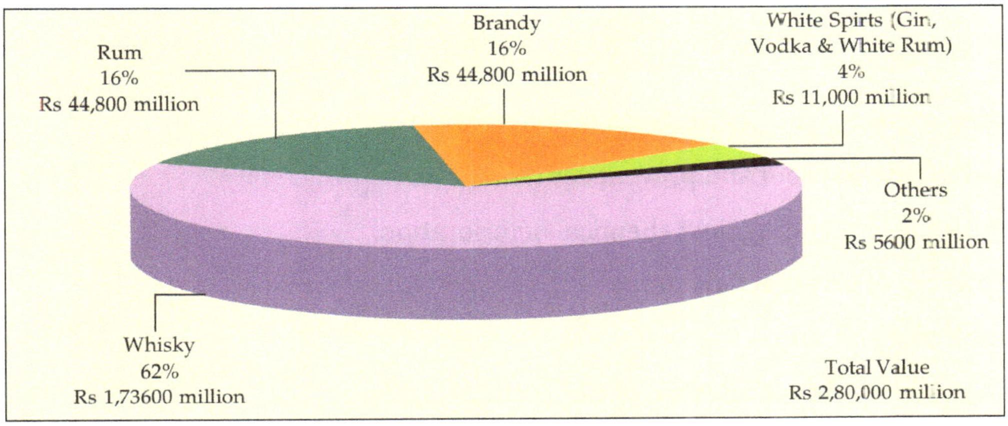

Market size under IMFL Category.

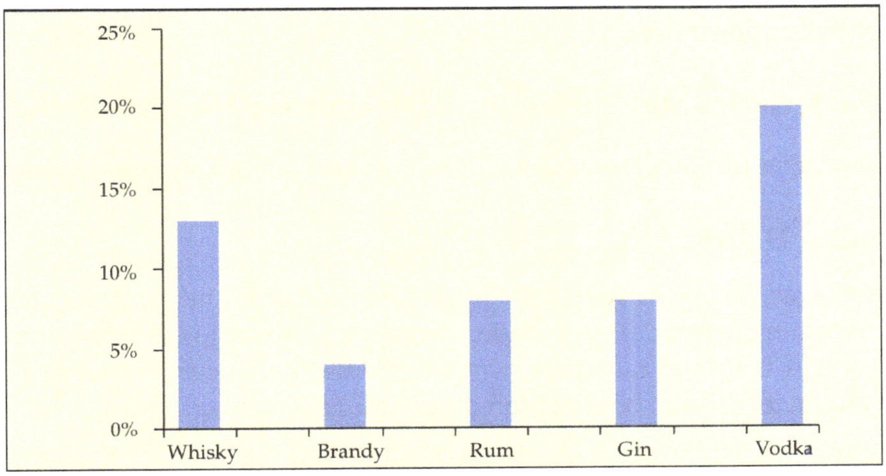

Growth Rate of IMFL Categories.

The term "beverage" is derived from French word "Beivre" which means a drink. General meaning of a drink includes a prepared drink. According to the Fruit Products Order (FPO) 1955 Act, Fruit Beverage or Fruits Drink means a beverage or drink which is prepared from fruit juice and water or carbonated water and containing sugar, dextrose, invert sugar or liquid glucose. The minimum percentage of fruit juice in the final product shall not be less than 5%. Fruit syrup connotes sweetened fruit juice of not less than 25% of fruit juice.

Classification

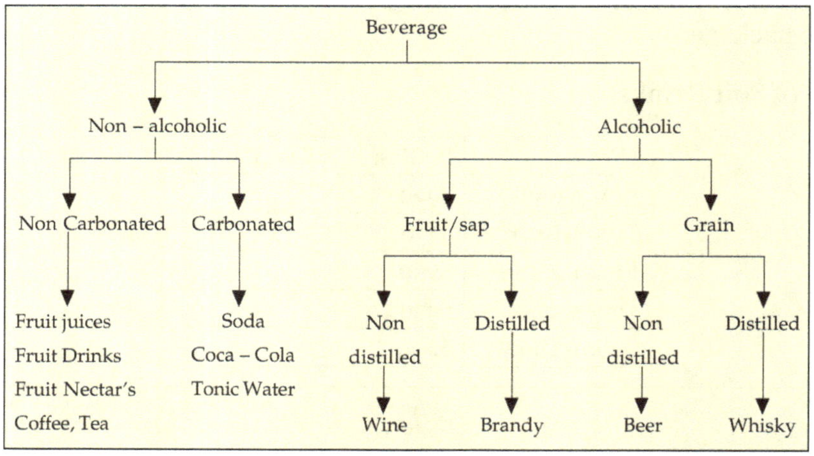

The packaging requirements for all types of beverages are:

- Absolutely leak-proof and prevent contamination.
- Protect the contents against chemical deterioration.
- No pick up of external flavours.
- Be hygienic and safe.
- Retain carbonation in the case of carbonated beverages.
- Economical, easy to use and dispose.
- Good aesthetic appearance.

Product Characteristics and Packaging Materials for Non-Alcoholic Beverages

Non-alcoholic beverages are further classified into non-carbonated and carbonated drinks.

Non-Carbonated Drinks

- Fruit Beverages: These include fruit juices, fruit drinks, and fruit nectars. They contain fruit juice, water, sugar, dextrose, invert sugar etc. The major deterioration that occurs in fruit beverages is loss of nutrition, physico-chemical changes and microbial growth. The product characteristics to be considered in relation to packaging are:
    - Acidity: All the fruit juices usually maintain an acidic character because they contain organic acids.
    - Enzymes: Enzymes exist in all fruit juices. Sometimes they have to be destroyed and some-times to be added.
    - Vitamin C (ascorbic acid): The Vitamin C content of a fruit increases until just before ripening, and then decreases due to the action of an enzyme, ascorbic acid oxidase. When fruits are cooked, most of the ascorbic acid transfers from the tissue into the liquid or may be oxidized, oxidation occurring more easily in iron, copper or

badly tinned vessels. Losses of Vitamin C also occur during storage. Storing at low temperature and preventing contact with air and light reduces this. Addition of sulphite has a preserving effect on Vitamin C.

- Colour and Flavour: They are very important, and many fruit drinks contain certain legally permitted colourings. These are added to overcome the bleaching effect of the sulphite used as a preservative, and to provide an attractive appearance.

Spoilage of fruit juices is mainly due to yeast. Its growth depends upon the temperature. Spoilage of raw fruit juice at room temperature results in alcoholic fermentation followed by the oxidation of alcohol and fruit acids by yeasts or moulds growing on the surface. Hence every living yeast cell must be removed or suppressed by pasteurization, filtration and/or preservatives.

Key parameters to be considered when selecting a packaging system are:

- Process.
- Distribution, shelf-life requirements, legislation.
- Product composition and quality as produced and at full shelf-life.
- Product protection required during storage, distribution and retail sale.
- Pack size, printing options, display etc.
- Packing system concept, automation options, ability to integrate with existing and/or future systems.
- Consumer appeal, image of product and packing.

The different packaging materials used are:

1. Glass Containers: The use of glass bottles for the packaging of fruit beverages was widespread although the hot-fill/hold/cool process had to be applied with care to avoid breakage of the containers. Glass is still the preferred packaging medium for high quality fruit beverages. However, over recent years, an increasing proportion is being packed aseptically, into cartons.

The improvements that have occurred in glass bottle packaging are:

- Light weight.
- Surface coating to increase abrasion resistance.
- Use of wide mouth containers fitted with easy-open-caps.

2. Metal Containers: Tinplate cans made of low carbon mild steel of 99.75% purity, coated with tin with easy open ends are used. These tinplate containers are either 3 piece or 2 piece containers. They are lacquered internally to prevent corrosion.

3. Plastic Containers: Fruit juices contain organic substances, which are sensitive to bacterial contamination. Packaging of such products is done through hot filling, to achieve extended shelf-life, PET bottles are usually used for hot filling applications. Special features are added to the containers through design and manufacturing process. The package is heat-set in order to improve

the temperature resistance of the containers. PET resins with a higher Tg (glass transition) temperature and/or a faster rate of crystallisation are used.

Normally hot-filled PET bottles are designed about 1.5 times heavier than cold-filled bottles. Reinforcing ribs and grooves are also provided along the circumference and base of the bottle. After filling and capping operations, the liquid continues to cool, which results in formation of vacuum. The bottle wall can deform under the influence of vacuum, and this problem is overcome by providing vacuum panels in the container side wall. Generally, lower levels of PET co-polymer are preferred and intrinsic viscosities of about 80 are acceptable.

Flexible plastic packages offer economic savings over conventional glass and metal containers but they are permeable to oxygen. Therefore, it is critical to select a flexible package that minimizes the permeability to oxygen. Flexible laminated pouches like metallised polyester/polyester/polyethylene are used for hot fill packaging method without retorting for acidic fruit juices. These are used either as flat pouches or stand-up pouches. However, the shelf-life of the product in these pouches is limited.

Fruit Drinks in Aseptic Packs.

4. Aseptic Packages: Ready to serve fruit beverages and fruit pulps/concentrates, packed in aseptic packages provide excellent protection for fruit juices/pulps. These aseptic packages are made by combining thermoplastic with paperboard and aluminium foil. Their multi-layered construction enables the carton to protect the contents from various factors responsible for spoilage. The aluminium foil layer is a strong barrier for O_2 and light. The inner plastic layer made of polyethylene makes it possible to seal through the liquid. The outer paper layer provides stiffness making it possible for the cartons in a brick shape, thus, enabling maximum utilisation of available storage and transportation space. Excellent graphics are possible leading to good display and shelf appeal and also providing information regarding the product. The aseptic process makes the product bacteria-free before being packaged.

To provide convenient access to the contents, beverage cartons offer a variety of opening devices. A familiar opening feature of the pack is the drinking straw, which is attached to the package. Some recent trends are pull-tab opening, which can be readily detached from a pre-punched hole without compromising the package integrity. Also, custom designed caps and closures can be incorporated on beverage cartons for easier pouring and for enhancing the brand image.

Also, the beverage cartons are now available in new prisma shape, which is comfortable to hold, and the unique shape offers maximum display effectiveness and high space efficiency. These packs are shelf-stable at room temperature and the shelf-life and nutrient composition of the fruit juice is influenced by the barrier properties of the tetrapak.

5. Bag-in-Box System: It consists of a collapsible bag within a rigid container, a filling machine to introduce the liquid product into the bag and a dispenser to draw the product out.

The outer container can be a box, a crate or a drum. The bag actually consists of two bags. An inner bag contains the liquid and an outer bag provides the barrier properties. Both are heat-sealed at the edges. The tubular spout fitted to the bag aids in filling and dispensing of the product. As little as 3 litres or as much as 1000 litres can be packed. The bag is the "life" of the system. The bag itself consists of three components:

- An inner layer,
- An outer layer,
- A spout.

The function of the inner layer, the one in contact with the material being packed, is to provide the bag with seal integrity. The seals are to be strong enough to withstand constant mechanical and chemical "pressure" for at least twice the expected shelf-life of the product. Generally, the inner layer is not designed for barrier unless the product needs extra barrier, which the outer layer cannot provide. Plastic films manufactured from high performance polyethylene, with excellent sealing and puncture properties are usually used as inner layer materials.

The function of the outer layer is to provide the bag with barrier commensurate with the expectation of the shelf-life of the product. In this respect, bag in box scores over other packaging forms, because unlike other packaging materials like jars and cans, the barrier property and hence the cost of this packaging form can be varied. The shelf-life expectations and storage condition play an important part in determining what the barrier requirements of the outer layer need to be. The standard outside layer is a metallised film laminate, which under standard conditions has an OTR of $1cc/m^2/24hrs$.

The spout and cap assembly are made of injection moulded plastics. The spout is provided with a flange, which is welded to the bag's inner layer during bag's manufacture. The spout and cap have two functions. They are used to fill the product in the bag and are also used to dispense the product from the pack. Several spout types are available ranging from a simple bung like configuration to ones, which can only be opened on the filling machine. The caps come with various tamper evident features as well.

6. Rigid Containers: The purpose of this container is to hold the bag during storage, transportation and use. As is evident, the bag being made from flexible films is incapable of being stored and transported by itself. Hence, each bag is individually stored into rigid outer container post filling. For packs of capacity below 50 litres, corrugated fibreboard cartons are generally used as the rigid containers. Beyond that plastics and metal drums are used. For very large bags, 1000 litres, + polygonal (6-8 faces) corrugated fibreboard containers made from 7 to 9 ply are used, though there are other alternatives as well.

Coffee

Coffee is made from the coffee beans, which is converted into a consumable beverage. Some of the major deterioration reactions in coffee are:

- Staling: This may be due to loss of flavour volatiles or due to chemical changes caused by moisture and oxygen absorption.
- Evolution of CO_2, which is emitted during the roasting process.
- Ingress of moisture in instant coffee results in caking. This usually happens when the moisture content reaches 7-8%. The initial moisture content of instant coffee is 2-4%.

Hence, while developing packaging system for coffee the following are to be considered:

- Moisture vapour ingress.
- Oxygen permeability.
- CO_2 and Volatile component egress.
- Grease resistance.

The packaging materials used are tinplate containers, composite containers, glass jars and flexible plastic pouches. The flexible laminates most widely used are 12μ PET/2μ Al Foil/70μ LDPE and MET PET/LDPE. Aluminium foil lined plastic pouches are most popular having 59% contribution in terms of volume.

Tea

The unique taste and colour of tea is because of polyphenols (cate chins) and amino acids (theamine). The flavour is due to the essential oils present in fresh leaves and volatile components developed during the manufacturing process. The different types of tea are: Black tea (fermented) Oolong tea (semi-processed) and Green tea (non-fermented). Deterioration is caused by loss of volatile components, increase in undesirable "taints" arising from oxidation reaction from fatty acids. Deterioration in green tea is caused by reduction in ascorbic acid content, change in colour from bright green to olive green and change in odour. The most common packaging material is paperboard carton with a liner or an overwrap of PP or regenerated cellulosic film. Other types are plastic jars, bottles, pouches, strips and envelopes. Plastic pouches have captured 12% of the tea market.

Carbonated Drinks

Carbonated drinks contain carbonated water, flavour, colour, sweeteners and preservatives. CO_2 gas from pure source is dissolved in water (amount varies with different types of beverages). A variety of ingredients like flavouring agents, colouring agents, preservatives, artificial sweeteners, antioxidants and foaming agents are then added.

Two major deteriorative changes that occur in carbonated drinks are the loss of carbonation and rancidification of essential flavouring oils. The first is largely a function of the effectiveness of the package in providing a barrier to gas permeation, while the latter can be prevented by the use of high quality flavourings and antioxidants, and de-aerating the mix prior to carbonation. Oxidative rancidity is reduced by the effectiveness of the package in providing a barrier to gas permeation.

Hence, the carbonated drink package requires a container that will hold pressure and not contribute off flavours. For many years virtually all carbonated soft drinks were packaged in glass bottles sealed with crown cork. In recent years, non-returnable glass bottles are giving way to refillable bottles. These have a foam plastic protective label of paper/poly or an all plastic shrink sleeve, as a safety measure to prevent flying of glass fragments in case of breakage of these containers. The crown closure has been replaced with a roll-on aluminium screw cap with tamper proof facility. Among the metal containers, the 3-piece tinplate containers have been used since long for the packaging of carbonated beverages. These are being replaced now by 2-piece aluminium cans. These cans retain the integrity of lacquer better than tin cans. Vinyl, epoxy and vinyl organosol coatings are used as lacquers for aluminium cans. Epoxy amine provides good adhesion, colour and flexibility to the can.

Among the plastic containers, PET bottles are the most preferred packaging material for packaging of soft drinks. The factors that influence the selection of a plastic package intended to contain CSD are indicated in Figure.

Factors to be Considered for CSD Packages.

Soft drinks have a maximum permissible level of 20ppm for citrus flavoured beverages and 40ppm for cola drinks while the water loss is of the order of 1%. Also, the loss of CO_2 through the wall must be allowed for. While increasing thickness will decrease the rate of CO_2 permeation, the cost of the bottle, will also increase and so a compromise has to be made. Other problems to be considered in plastic containers are creep and elastic deformation. The polyethylene terephthalate (PET) bottle satisfies most of the requirements for packaging of carbonated soft drinks. Improved blow moulding techniques and bi-axial stretching have made PET container to be pressurised due to its strength, dimensional stability and precision. Also, they have a glass like appearance, good transparency, lustre, chemical inertness and unbreakability.

Advantages of PET Containers

The advantages of PET container are:

- Superior packaging to product ratio: PET container being 63% and 47% more energy efficient than glass bottles and aluminium cans respectively.
- PET bottles are 32% more energy efficient than glass bottles during delivery of 1000 gallons of soft drinks.

- Glass bottles and Aluminium-cans generate 230% and 175% times more atmospheric emissions compared to PET.
- PET bottles contribute 68% and 18% less solid waste by weight compared to glass and aluminium containers.
- 100 kg of oil is required to produce 1000 1-litre PET bottles as against 230 kg for 1000 equivalent glass bottles.
- PET bottles help in fuel saving due to their lower weight.

The resins used in PET bottles to pack carbonated drinks are of a very special quality. The PET bottles have to be extremely strong to contain the internal pressure of CO_2 without distortion and expansion. This is obtained by using a resin, which has high intrinsic viscosity and lower co-polymer levels.

Currently, more than 90% of PET is consumed in food packaging with beverages/drinks forming nearly 80%. The breakup of world PET resin consumption by weight expressed as percentage share is shown in the Figure.

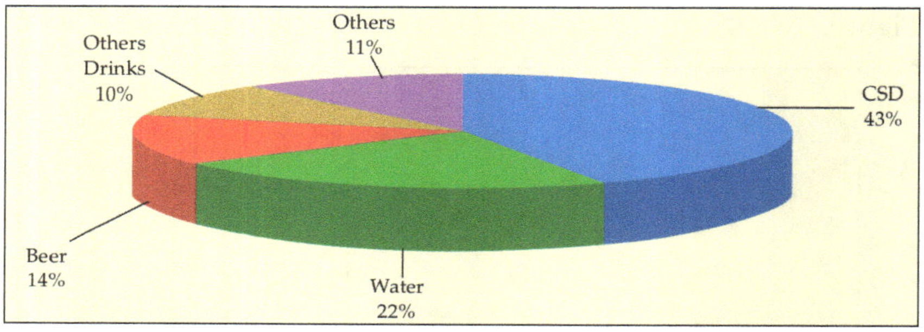

PET Resin Consumption.

Among other materials suitable to contain CSD, polyethylene naphthalate (PEN) meets several requirements such as physical, chemical and barrier properties. It has a barrier to CO_2, O_2, water vapour and UV light which is 4-5 times better than PET. Also, PEN is lighter in weight and has better shatter resistance. As pure PEN is expensive, it is blended with PET to form a co-polymer, still far too expensive to be used by bottle manufacturers.

Product Characteristics and Packaging Materials for Alcoholic Beverages

Alcoholic Beverages in PET Bottles.

Alcoholic drinks originated through the action of yeast cells on sugar containing liquids. Alcoholic drinks are aromatic liquids with specified alcohol content. Some kinds contain carbon dioxide, others a quantity of sugar. They are either fruit/sap based or grain based. They can either be non-distilled or distilled depending on the volume percentage of alcohol per litre. The border between the two kinds of drink is about 20%. The different types of alcoholic beverages are beer, wine, whiskey, brandy etc.

Non Distilled Alcoholic Beverages

1. Beer (Grain Based): Beer is made from grains and has low alcoholic content around 5% by volume. Barley is the chief grain, but rice and corn are also used. The grains are brewed and fermented and then carbonated with CO_2 and flavoured with hops to give a bitter flavour.

Owing to its low pH (about 4.0), microbial degradation is not usually a problem with beer, and the use of pasteurization and aseptic cold filtration excludes yeast. However, during storage beer can undergo irreversible changes leading to appearance of haze, development of off-flavours and increased colour. The oxidation reaction gives beer a "card-board-like" flavour. Flavour loss is also accelerated in the presence of light and certain metal ions. The fermentation process consumes oxygen. Also, brewing reduces the level of oxygen in beer to 40-50 ppb prior to packaging.

During the packaging process, atmospheric oxygen enters the package and the level of oxygen contamination reaches 250-500 ppb, which corresponds to 0.1-0.2 ml of oxygen per 335 ml bottle or can. This results in a shelf-life of beer of 80 to 120 days. The oxygen consumption of beer varies with the composition of the beer, its age, presence of reducing agents, temperature etc.

The traditional packaging media for beer is the glass bottle sealed with a crown closure. Recent development is the use of PET bottles for packaging of beer. Types of PET beer bottles used are non-tunnel pasterurised, one way tunnel pasterurised and returnable/re-fillable bottles.

Beer needs high performance in both CO_2 and O_2 barrier compared to PET used in carbonated soft drinks (CSD) applications. The level required depends on the type of beer, container size, distribution channels and environmental conditions (storage time, temperature and humidity levels). Improvements in the barrier can be obtained via colourants, creating multi-layer bottles and scavengers. Protection from U. V. light is obtained by adding colourants or U. V. additives, during the injection moulding stage. Because of the varied requirements for beer, the resins chosen must provide an adequate barrier, UV protection and clarity. Beer bottles need strength in order to maintain the CO_2 pressure over a wide range of temperatures. Intrinsic viscosities in the range of 0.8 to 0.84 are normally used. Since the PET bottles are lighter, a truck can carry 60% more of the beverage and 80% less packaging–a fuel saving of 40% and less air pollution.

2. Wine (Fruit/Sap based): Wine is a beverage resulting from the fermentation by yeasts of the juice of grapes with appropriate processing and additions. The major deteriorative reaction in wines is caused by oxidation, the oxygen gradually changing the wine character, leading to development of browning and undesirable flavours.

The most common form of packaging used for wines is the glass bottle sealed with natural cork. Since wines are affected by sunlight, the bottles usually used are of coloured glass. Bottled wine is normally stored in the horizontal position so that the cork is kept moist, thereby providing a better barrier to the ingress of oxygen.

Alcohol Beverages in PET Bottles.

The most significant change in the packaging of wine resulted from the development of the bag-in-box package: a flexible, collapsible, fully sealed bag made from one or more plies of synthetic films, a closure and a tubular spout through which the contents are filled and dispensed, and a rigid outer box or container. The bag is generally constructed from co-extruded film of EVA-BA-EVOH-BAE-VA or LDPE-BA-EVOH-BA-LDPE.

The physical strength of the bag is of prime importance and must remain intact throughout distribution and subsequent storage. Under normal circumstances, the bags are subjected to two forms of stress: hydraulic shock (normally caused by sudden acceleration/deacceleration of the pack) and flex crack. By using polymers, which have high flex resistance and improving the adhesion between the films, the strength of the bag can be increased. One problem associated with the packaging of wine into bag-in-box system is the decrease in shelf-life as compared to that obtained using traditional glass bottles. This is due to permeation of oxygen through the valve material of the tap. By improving the barrier properties of the bag and the design of the tap the problem can be solved. A special heatsealable membrane is partially attached to the gland during manufacture of the bag. After filling but prior to insertion of the tap, this membrane is fully sealed to the gland and the potential entry path is greatly reduced. As a result the bag is completely sealed and is totally independent of the tap until the membrane is broached when the pack is just opened by the consumer. A cross section of the membrane and gland before and after filling is shown in Figure. Wines are also available in PET bottles and in stand-up pouches of metallised polyester laminates.

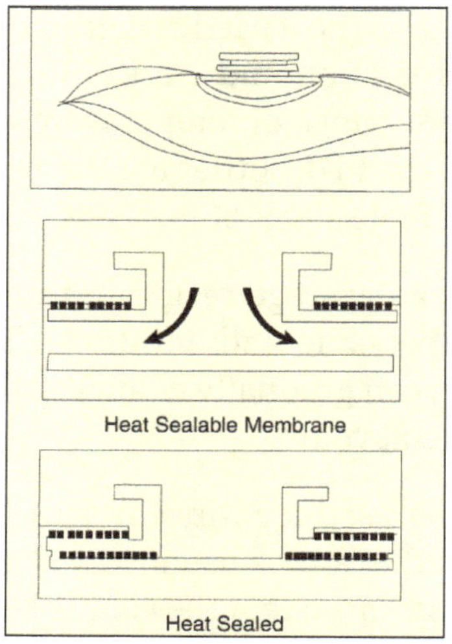

Bag-in-box before and after Filling.

Distilled Alcoholic Beverages

Brandy and whisky are these drinks are obtained by distillation of alcohol containing drinks. During distillation the aqueous part is separated from the alcohol. The distillates obtained are sold under several names like brandy, gin, whisky cognac, vodka, etc and have different alcohol percentage.

Because of their high alcohol percentages, these liquors are mostly packed in glass bottles so that they can be kept for an infinite time after opening. The bottles are sealed to prevent alcohol from evaporating and to protect the contents of the bottles from dirt and dust.

References

- Materials-used-in-food-packaging: matmatch.com, Retrieved 17, February 2020
- Trends-in-Fruit-and-Vegetable-Packaging-210280254: researchgate.net, Retrieved 10, May 2020
- Beverages, icpefoodnpackaging: icpe.in, Retrieved 23, August 2020

Chapter 4

Diverse Aspects of Food Engineering

Food engineering entails the study and applications of various concepts and devices such as biosensors, food adulteration and non-destructive food analysis. Biosensors find a variety of uses in food engineering, especially in detecting pathogens and food toxins. The topics elaborated in this chapter will help in gaining a better perspective about these aspects of food engineering.

Biosensors

A biosensor is an analytical device, used for the detection of an analyte that combines a biological component with a physicochemical detector the sensitive biological element (e.g. tissue, microorganisms, organelles, cell receptors, enzymes, antibodies, nucleic acids, etc.), a biologically derived material or biomimetic component that interacts (binds or recognizes) the analyte under study. The biologically sensitive elements can also be created by biological engineering.

- The transducer or the detector element (works in a physicochemical way; optical, piezoelectric, electrochemical, etc.) that transforms the signal resulting from the interaction of the analyte with the biological element into another signal (i.e., transduces) that can be more easily measured and quantified.
- Biosensor reader device with the associated electronics or signal processors that are primarily responsible for the display of the results in a user-friendly way. This sometimes accounts for the most expensive part of the sensor device, however it is possible to generate a user friendly display that includes transducer and sensitive element.

The Biosensor System

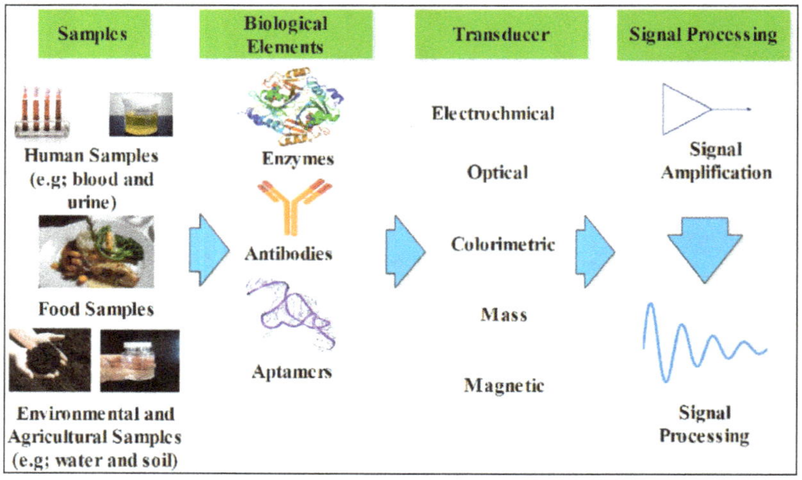

A biosensor typically consists of a bio-recognition component, biotransducer component, and electronic system which include a signal amplifier, processor, and display. Transducers and electronics can be combined, e.g., in CMOS-based microsensor systems. The recognition component, often called a bioreceptor, uses biomolecules from organisms or receptors modeled after biological systems to interact with the analyte of interest. This interaction is measured by the biotranducer which outputs a measurable signal proportional to the presence of the target analyte in the sample. The general aim of the design of a biosensor is to enable quick, convenient testing at the point of concern or care where the sample was procured.

Bioreceptor

In a biosensor, the bioreceptor is designed to interact with the specific analyte of interest to produce an effect measurable by the transducer. High selectivity for the analyte among a matrix of other chemical or biological components is a key requirement of the bioreceptor. While the type of biomolecule used can vary widely, biosensors can be classified according to common type's bioreceptor interactions involving: anitbody/antigen, enzymes, nucleic acids/DNA, cellular structures/cells, or biomimetic materials.

Types of Biosensors

Biosensors can be grouped according to their biological element or their transduction e includes enzymes, antibodies, micro-organisms, biological tissue, and organelles. Antibody-based biosensors are also called immunosensors. When the binding of the sensing element and the analyte is the detected event, the instrument is described as an affinity sensor. When the interaction between the biological element and the analyte is accompanied or followed by a chemical change in which the concentration of one of the substrates or products is measured the instrument is described as a metabolism sensor. Finally, when the signal is produced after binding the analyte without chemically changing it but by converting an auxiliary substrate, the biosensor is called a catalytic sensor.

The method of transduction depends on the type of physicochemical change resulting from the sensing event. Often, an important ancillary part of a biosensor is a membrane that covers the biological sensing element and has the main functions of selective permeation and diffusion control of analyte, protection against mechanical stresses, and support for the biological element.

Enzymes

Enzymes are proteins with high catalytic activity and selectivity towards substrates. They have been used for decades to assay the concentration of diverse analytes. Their commercial availability at high purity levels makes them very attractive for mass production of enzyme sensors. Their main limitations are that pH, ionic strength, chemical inhibitors, and temperature affect their activity. Most enzymes lose their activity when exposed to temperatures above 608C. Most of the enzymes used in biosensor fabrication are oxidases that consume dissolved oxygen and produce hydrogen peroxide. Enzymes have been immobilized at the surface of the transducer by adsorption, covalent attachment, and entrapment in a gel or an electrochemically generated polymer, in filliped membranes or in solution behind a selective membrane.

Enzymes are commonly coupled to electrochemical and fiber optic transducer Antibodies is proteins that show outstanding selectivity. They are produced by blymphocytes in response to antigenic structures, that is, substances foreign to the organism. Molecules larger than about 10 kDa can stimulate an immune response. Smaller molecules like vitamins or steroids can be antigenic (also called haptens) but they do not cause an immune response unless they are conjugated to larger ones like bovine serum albumin. Many antibodies are commercially available and commonly used in immunoassays. Antibodies are usually immobilized on the surface of the transducer by covalent attachment by conjugation of amino, carboxyl, aldehyde, or sulfhydryl groups. The surface of the transducer must be previously functionalized with an amino, carboxyl, hydroxyl, or other group. A review of conjugation techniques can be found elsewhere. Similar limitations with enzymes. Furthermore, binding may not be reversible and regeneration of the surface may require drastic changes in conditions like low pH, high ionic strength, detergents, etc. Therefore, efforts are being made to produce low cost, single use sensors. Probably the main potential advantage of traditional immunoassays is that they could allow faster and in-field measurements.

Microbes

The use micro-organisms as biological elements in biosensors is based on the measurement of their metabolism, in many cases accompanied by the consumption of oxygen or carbon dioxide, and is, in most cases, measured electrochemically. Microbial cells have the advantage of being can be more stable, and can carry out several complex reactions involving enzymes and cofactors. Conversely, they are less selective than enzymes; they have longer response and recovery times and may require more frequent calibration. Micro-organisms have been immobilized, for example, in nylon nets, cellulose nitrate membranes, or acetyl cellulose.

Transducer Elements

Electrochemical, Amperometric and potentiometric transducers are the most commonly used electrochemical transducers. In Amperometric transducers, the potential between the two electrodes is set and the current produced by the oxidation or reduction of electro active species is measured and correlated to the concentration of the analyte of interest. Most electrodes are made of metals like platinum, gold, silver, and stainless steel, or carbon-based materials that are inert at the potentials at which the electrochemical reaction takes place.

However, because some species react at potentials where other species are present, either a selective membrane is used or an electron mediator that reacts at lower potential is incorporated into the immobilization matrix or to the sample containing the analyte.

Potentiometric transducers measure the cheaper than enzymes or antibodies, potential of electrochemical cells with very low current. Field effect transistors (FET) are potentiometric devices based on the measurement of potential at an insulator– electrolyte interface. The metal gate of a FET can be substituted by an ion selective membrane to make a Ph transducer (pH ISFET). Enzymes have been immobilized on the surface of such pH ISFET to produce enzyme sensitized field effect transistors (ENFET).

Optical Fiber, optic probes on the tip of which enzymes and dyes (often fluorescent) have been co-immobilized are used. These probes consist of at least two fibers. One is connected to a light source of a given wave length range that produces the excitation wave. The other, connected to a photodiode, detects the change in optical density at the appropriate wavelength Surface Plasmon resonance transducers, which measure minute changes in refractive index at and near the surface of the sensing element, have been proposed. Surface plasmon resonance (SPR) transducers have been proposed. SPR measurement is based on the detection of the attenuated total reflection of light in a prism with one side coated with a metal. When a p-polarized incident light passes through the prism and strikes the metal at an adequate angle, it induces a resonant charge wave at the metal/dielectric interface that propagates a few microns.

The total reflection is measured with a photodetector, as a function of the incident angle. For example, when an antigen binds to an antibody that is immobilized on the exposed surface of the metal the measured reflectivity increases. This increase in reflectivity can then be correlated to the concentration of antigen.

Calorimetric transducers measure the heat of a biochemical reaction at the sensing element. These devices can be classified according to the way heat is transferred. Isothermal calorimeters maintain the reaction cell at constant temperature using Joule heating or Peltier cooling and the amount of energy required is measured. Heat conduction calorimeters measure the temperature difference between the reaction vessel and an isothermal heat sink surrounding it. Using highly conducting materials ensure quick heat transferred between the reaction cell and the heat sink. Finally, the most commonly used is the isoperibol calorimeter that also measures the temperature difference between the reaction cell and an isothermal jacket surrounding it. However, in this case the reaction cell is thermally insulated (adiabatic). This calorimeter has the advantage of being easily coupled to flow injection analysis systems.

Applications

One of the major driving forces for the development of biosensors is biomedical diagnosis. The most popular example is glucose oxidase-based sensor used by individuals suffering from diabetes to monitor glucose levels in blood. Biosensors have found also potential applications in the agricultural and food industries. However, very few biosensors have been commercialized.

Agricultural Industry

Enzyme biosensors based on the inhibition of cholinesterases have been used to detect traces of organophosphates and carbamates from pesticides. Selective and sensitive microbial sensors for measurement of ammonia and methane have been studied. However, the only commercially available biosensors for wastewater quality control are biological oxygen demand (BOD) analyzers based on micro-organisms like the bacteria Rhodococcus erythropolis immobilized in collagen or polyacrylamide.

Food Industry

Biosensors for the measurement of carbohydrates, alcohols, and acids are commercially available. These instruments are mostly used in quality assurance laboratories or at best, on-line coupled is produced and measured b line through a flow injection analysis system. Their implementation in-line is limited by the need of sterility, frequent calibration, analyte dilution, etc.

Potential applications of enzyme based biosensors to food quality control include measurement of amino acids, amines, amides, heterocyclic compounds, carbohydrates, carboxylic acids, gases, cofactors, inorganic ions, alcohols, and phenols. Biosensors can be used in industries such as wine beer, yogurt, soft drinks producers. Immunosensors have important potential in ensuring food safety by detecting pathogenic organisms in fresh meat, poultry, or fish.

Current Research and Trends

Because in many cases the transduction technology is well established, most of the research is focused on improving immobilization techniques of the biological element to increase sensitivity, selectivity, and stability. While critical, the latter has received relatively little attention probably in part because there is a tendency to design disposable devices that are most useful in quality assurance laboratories but do not allow on-line implementation for process control.

Another dynamic area of research is miniaturization of sensors and flow systems. Development of these technologies is mainly driven by the need for in vivo applications for medical diagnosis and may not find immediate use in the agricultural and food industries. After almost 40 yr of research in biosensors, a wide gap between research and application is evident. The lack of validation, standardization, and certification of biosensors has resulted in a very slow transfer of technology. With faster computers and automated systems this process should accelerate in the future.

Food Adulteration

Adulteration refers to a set of circumstances where a food product, when studied and analysed, fails to meet the required legal benchmarks. The methodology of adulteration is varied and may serve various objectives.

It may be aimed at increasing the grammage of the final product with a stoic indifference towards the content and final quality of the product. Substances may be added which are either available food items of sub-quality or non-food items. The horrifying reality in the case of meat adulteration is that among meat and meat products, components used to adulterate are water or ice, carcasses, or carcasses of animals other than the animal meant to be consumed. There is a blatant disregard for aesthetics and finesse also in this methodology.

Little wonder that there are recorded doctrines to state that in ancient Rome, public hanging was the immediate punishment meted out to anyone indulging in what they termed,"a sin next to adultery", the linguistic similarity between the two words is both evident and obvious and certainly not a coincidence.

There was a practice of actually offering food to fire, to understand its constituents and their purity. Furthermore, the offering of food to birds and animals in the Gurukul environment also served the same purpose of assessment of its efficacy, as animals are said to be more astute sensorily.

The whole concept of food adulteration has been assigned to serious and protracted legal acts and one of the first objectives is to define accurately, the concept of adulteration. The Federal Food,

Drug, and Cosmetic (FD&C) Act 1938 specifies that food is "adulterated" if it meets falls in any of the following categories:

- Detrimental to health because of poisonous or deleterious content.
- Content bearing a substance cited elsewhere in the Act as unsafe.
- Container Composition being unsafe.
- Product bears or contains a pesticide chemical residue that is unsafe.
- It is, or it bears or contains, an unsafe food additive or stabilizer.
- It is, or it bears or contains an unsafe new animal drug.
- It is, or it bears or contains an unsafe colour component.
- It consists, in whole or in part, of some filthy, putrid, or decomposed substance" or is otherwise unfit for food.

It has been prepared, packed, or held under unsanitary conditions (insect, rodent, or bird infestation) whereby it may have become contaminated with filth or rendered injurious to health. Other corollaries of adulteration also include irradiation, bulking or indiscriminate increase of weight of the food product with a substandard analogue. Importantly, mislabelling, misbranding and even deliberately provided inaccurate information of either the compositional detail or the shelf life details are emphatically regarded as adulteration. It would be worth noting that the foods most adulterated are milk and milk products, meat, eggs, vegetables and fruits. The foods which are subjected to adulteration maybe both fresh and processed. Interestingly, there are very simple and scientific tests done to assess and detect adulterants. They do not even need instruments and many can be done at home.

They are commonly known as 'field tests'...because they can be conducted in situ and in the field/ area of food production. They are also referred to as 'platform tests' because they can be easily executed in the laboratory or home on a work surface or platform. It would be a worthwhile exercise to consider some common adulterants of each of the foods mentioned above and a simple test or two to detect them.

1. Milk: Milk is ironically the easiest food to adulterate. It is consumed in large quantities world over and is regarded as a "paediatric to geriatric food" and hence, there is an unscrupulous temptation to undermine its quality to increase their profit margins and sales bandwidth.

Milk can be adulterated by urea, starch, sugar, more callously by soda bicarb and even different flours. To increase volumes, water is used and the quality of water would have a definite say on the resultant water. Instruments like Lactometer are commonly used to grade the quality of milk. Tests using iodine for starch, resorcinol for cane sugar or sucrose and even a simple clot on boiling could give details of type or amount of adulteration. In the premise of human health and safety, the procedures and protocols for all these tests are Global Access Documents, much available to one and all for use and propagation. Similar tests are used for the assessment of milk products also.

2. Meat: Meat adulteration has both health and aesthetic repercussions. Significantly, type of meat consumption is categorized as per religious and social norms and since it is possible to mask the

meat cuts by use of colour and spices, it is possible to adulterate meat cuts with the meat of other animals. Often these animals are beyond the realms of acceptance. Machines and instruments like tenderometer and shear are used to measure tenderness and freshness of meat. Moreover, the animals are sacrificed as is where is and need-based so that carcasses and stale meat do not find their way to the consumer.

3. Eggs: In the case of this very versatile and popular food, it is possible to adulterate the food in two manners. If the poultry is fed with contaminated feed and feed doctored with hormones and steroids, it results in eggs which may be contrary to safety and wellness norms. Additionally egg products like desiccated egg powders and egg dishes could be prepared using eggs other than hen eggs. Misbranding and mislabelling of ingredients are aimed at abusing the trust of the consumer.

4. Vegetables and fruits: Freshness and hydrated texture and colour are desirable attributes for both vegetables and fruits hence, adulteration aims at presenting the natural produce falsely and misleadingly to the consumer.

Spraying of water laced with chemicals of indifferent nature and colouring material which may be potentially allergy-causing properties or which may even be carcinogenic (cancer-causing). Vigorous washing, personal handling of the farm produce to judge the physical and sensory characteristics of them can help.

The underbelly of food processing is both a reality and something that cannot be underestimated or ignored. Adulteration undermines our health and wellness through no fault of ours. It is important to recognise it as a crime in all its evil dimensions and treated as such.

Non-destructive Food Analysis

Quality assessment of foods is essential for food research and development .chemical and physical analyses are conducted by using various chemicals and instruments to asses' food Quality and the analyses are usually time consuming and costily. When a sample has been used for chemical or physical analysis, it usually cannot be used for other analyses.

Quality and safety is the key factor in modern food industry. Currently conventional food measurement methods are destructive and inefficient; therefore development of non-destructive and efficient measurement tool is important.

Recently, optical sensing technologies have been investigated as potential tools for non-destructive evaluation and inspection for food quality and safety. In particularly, hyper spectral imaging (or imaging spectroscopy) which is based on two mature technologies of imaging and spectroscopy have been widely studied and developed, resulting in many successful applications in the food industry.

Principles of Non-destructive Food Analysis

A conventional imaging system or more specifically computer vision is a common technique for obtaining spatial information of the sample in the forms of monochromatic or colour images, therefore conventional imaging system is used for colour, shape, size, surface texture evaluation of food products and for surface defects detection in food inspection, however it cannot identify or detect chemical properties or characteristics from a food product.

On the other hand, conventional spectroscopy system is a technique for evaluating chemical properties or characteristics of food products. However, conventional spectroscopy system cannot cover a large area or a small area with high spatial resolution. Therefore the technique does not provide the spatial information which is regularly required for and is critical in food inspection.

With the integration of the main features of imaging and spectroscopy, hyper spectral imaging can simultaneously acquire both spatial and spectral information that is critical to the detection of food safety and evaluation of food quality attributes. A typical hyper spectral system consists of a light source, a wavelength dispersion device, and an area detector. The images are acquired over the visible and near-infrared (or infrared) wavelengths to specify the complete wavelength spectrum of a sample at each point in the imaging plane. These images are then combined and form a three dimensional hyper spectral cube, with two dimensions for describing spatial information and the third one for spectral information. In this hypercube, each spectral pixel corresponds to a spectral signature (or spectrum) of the corresponding spatial region, recording the entire measured spectrum of the imaged spatial point. Therefore the measured spectrum indicates the ability of the sample in absorbing or scattering the exciting light, representing the inherent chemical properties of a sample. As a result, the technology provides us with unprecedented detection capabilities, which otherwise cannot be achieved with either imaging or spectroscopy alone. If conventional imaging is to provide the answer to the question of where and conventional spectroscopy is to provide the answer to the question of what, hyper spectral imaging is a technique to provide the answer to the question of where is what.

Hyper spectral imaging techniques have received much attention for food quality and safety evaluation and inspection. Many approaches and applications have shown the usefulness of hyper spectral imaging in the food industry.

Applications of Non-Destructive Food Analysis

Poultry Carcass Wholesomeness

The pioneer work of using hyper spectral imaging technique for automated poultry carcass inspection is conducted by USDA Agricultural Research Service (ARS). Since 1960s; USDA ARS has been researching on spectroscopy and spectral imaging methods for non-destructive food quality and safety measurement. Their current ongoing studies includes the development of automated spectral imaging for detecting the surface contaminants on fresh produce and for inspecting wholesomeness and detecting contamination on poultry carcasses, all on high-speed processing lines. Their continuing efforts have led to the development of the latest hyper spectral line-scan imaging system for wholesomeness inspection on commercial high-speed processing lines, which is now under commercial development for industrial use. In addition, USDA ARS has also carried out fundamental research for developing feasible spectral methods to assess poultry characteristics, and adapting the scientific findings to implement in the current ARS imaging system for practical real-world use in automated high-speed processing environments.

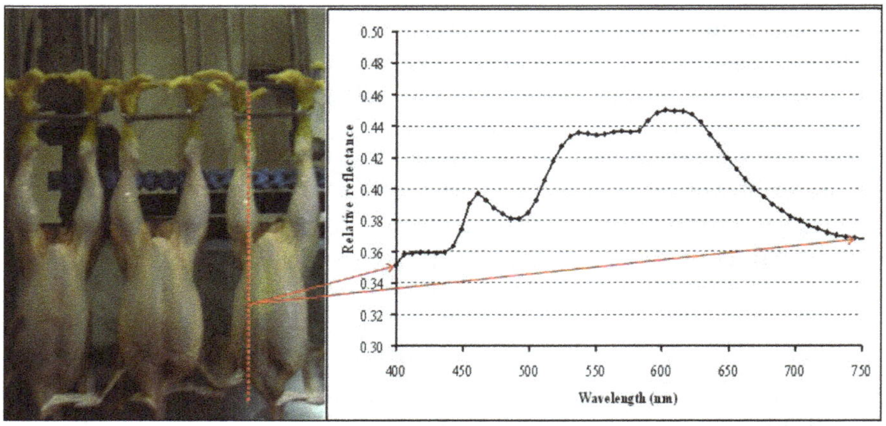

Full-spectrum data are acquired for every pixel in each hyperspectral line-scan image.

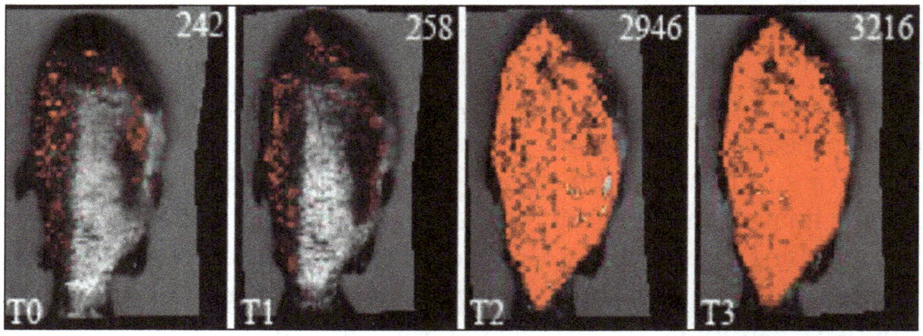

Example of fish freshness classification (pixels in red are those classified as "non-fresh").

Fish Freshness

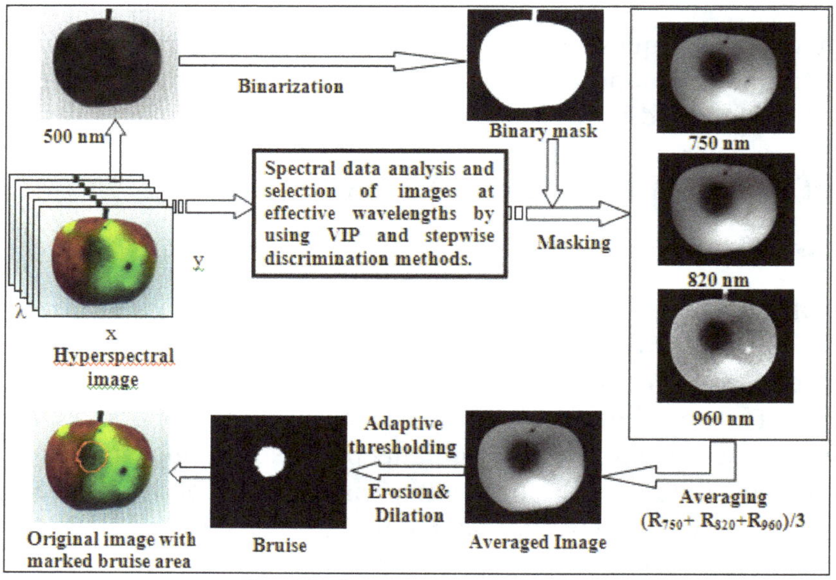

An example of the key steps involved in apple bruise detection algorithm using hyperspectral imaging technique.

In recent years, hyper spectral imaging has been used in fish meat quality tests, including salmon, trout, cod, halibut, and sea bass. Most of these applications are related to chemical composition prediction, protein contents, and levels of moisture contents. However research has been performed to use hyper spectral imaging as a method to provide an objective and qualitative evaluation of the state of the fish freshness. The study focused on establishing a correlation between the spectral reflectance of selected areas of the epidermis and the time of storage in standard refrigeration procedures, and the possibility of finding objective parameters for the good prediction of fish freshness that considers products stored for more than three days as "non-fresh but still edible". Through this research, it has been demonstrated that it is possible that hyper spectral imaging will provide a valid contribution in relation to the monitoring of the organolepctic and mercelogic properties of fish production during all steps along the production chain.

Bruise Detection of Apples

The bruise damages of apples are normally due to impact, compression, vibration, or abrasion during handling. The impact bruise may not be visible immediately when the impact applies. The symptom appears after a certain period of time. Therefore early detection of such impact bruise is needed in order to improve the product quality. One of the methods is to use the hyperspectral imaging technology. Many studies have been carried out on the area which shows the potentials of using the technique on detecting apple bruises effectively. However, the speed, cost, and processing power required make it still be serving more on research than practical applications. However with the development of new spectral imaging systems with lower cost, wider spectral range, and better dynamic range, the technique will find new and boarder arena of practical applications in the fruit industry.

Citrus Fruit Inspection

Citrus fruit is another type of fruits that require early detection due to the fact that a small number of fruit that infected by fungi can spread the infection to a whole consignment of citrus fruit. This problem is even more severe if the fruit is stored for a long period of time or during long term transportation for exportation.

As hyper spectral image technique allows studying the reflectance of defects and other regions of interest in particular wavelengths, it has been investigated as an important tool for the quality inspection of citrus fruits, offering the possibility of designing machines for the automatic identification of blemishes. which is particularly important for early rot detection. RGB =550 nm =660 nm =950 nm.

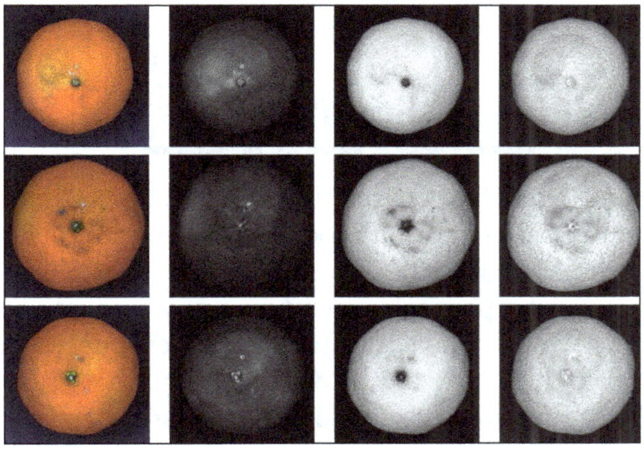

RGB and monochromatic images (550 nm, 660 nm and 950 nm) of various mandarins.

Sugar Distribution of Melons

One of the very interesting applications of hyper spectral imaging technique is to use it to predict the sugar content distribution in melons. In this research, the relationship between the sugar content and absorption spectra was investigated, and it was found that the absorbance at 676 nm was close to the absorption band of chlorophyll and exhibited a strong inverse correlation with the sugar content. A calibration method was used to convert the absorbance values on the image into the Brix sugar content in accordance with the spectroscopy techniques. When each pixel of the absorption image was converted, a colour distribution map of the sugar content could then be constructed as shown in figure.

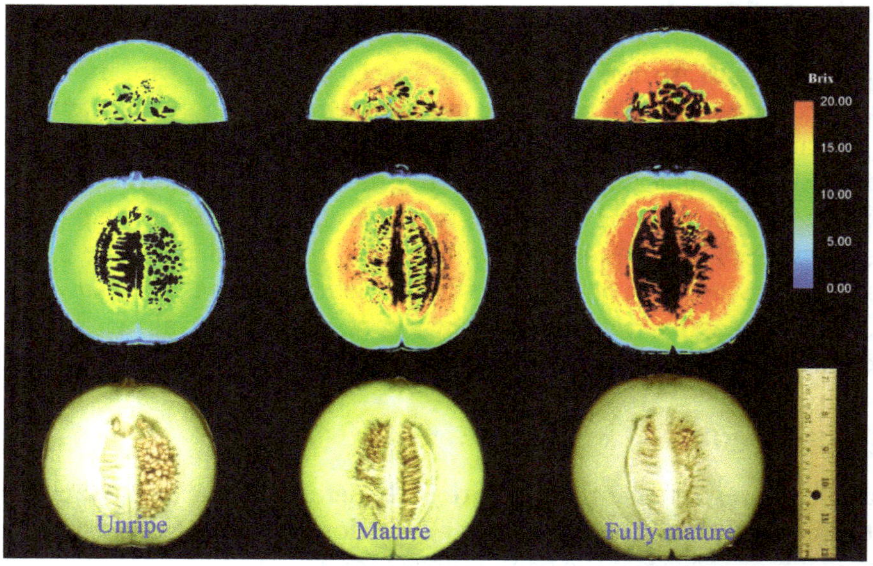

Sugar distribution map for unripe, mature and fully mature melons.

Other Applications

Hyper spectral imaging technique has been studied for its potentials in quality control and analysis for many other foods. These include meat quality assessment measuring ripening of tomatoes quality evaluation of mushroom defect detection of pickling cucumber and classification of wheat kernels.

Membrane Separation Technology

Among the wide range of the available separation processes, membrane technology is defined as abroad term that contains several separation processes on molecular level i.e. membrane separation usually applied on10 μm size molecules. It is among the vital fields of food science. A set of specialized equipment is used for a particular membrane process that enable it to show its typical separation characteristics as well as, make it suitable for specific desired single or multiple applications. Water industry (drinking water + wastewater processing + water desalination) has the largest commercial market for membranes worldwide. In last 20-25 years, the market volume of different membranes has increased about€800–850 million in food industries and the same is now stands as the second largest commercial market for membrane on global basis. Moreover, about 20-30 % of the total membrane production is used in food industries, annually increasing at the rate of 7.5% in this sector on global basis. Membrane separation is performed above the atmospheric pressure (that varies with particular membrane process) in a closed system so, these processes are known as pressure-driven membrane processes. Four membrane processes (in their ascending pore diameter) are Reverse osmosis (RO), Nanofiltration (NF), Ultrafiltration (UF) and Microfiltration (MF) are mainly used in food processing. UF systems as well as UF membranes have the maximum share (35%) in global membranes market followed by the share of MF systems and membranes (33%) and NF/RO systems and membranes (30%). Membrane contactors (MC), Electrodialysis (ED) and Pervaporation (PV) are the other membrane processes but contribute only for 2% share in total membranes market.

Classical Membranes, their Types and Design

Membranes are semi-permeable barriers, used to separate two phases, which restrict the transport of various substances in permeate and retentate streams in a specific way i.e. on the basis of particle size, shape, electric charge etc. The separation efficiency of a particular membrane is usually affected by different processing factors like feed composition, pH, temperature, pressure, feed flow as well as interactions between feed component and membrane surface. During membrane processing, a particular membrane material is fixed in a particular membrane module (Plate & frame, Spiral wound, Hollow fiber, Tubular) and operated within pre-fixed operating conditions to perform assigned tasks. Membrane processes were initially operated in dead end filtration mode in which feed (single stream) was subjected to membrane that retains the bigger size molecules in retentate and allowed the smaller molecules to pass in permeate (single stream). The main problem of such systems is the rapid cake layer formation over the membrane surface which is directly responsible for membrane flux decline. This problem was overcome in cross flow design in which single feed stream is subjected to membrane, parallel to its surface in closed system at higher pressure with turbulent flow that divide the feed in to two streams namely permeate (molecules that passed through membrane) and retentate (containing rejected molecules).The turbulent flow helps in reduction of cake layer formation through continuous sweeping action of suspended particles over the membrane surface that results in better flux than dead end filtration over the higher length of time.

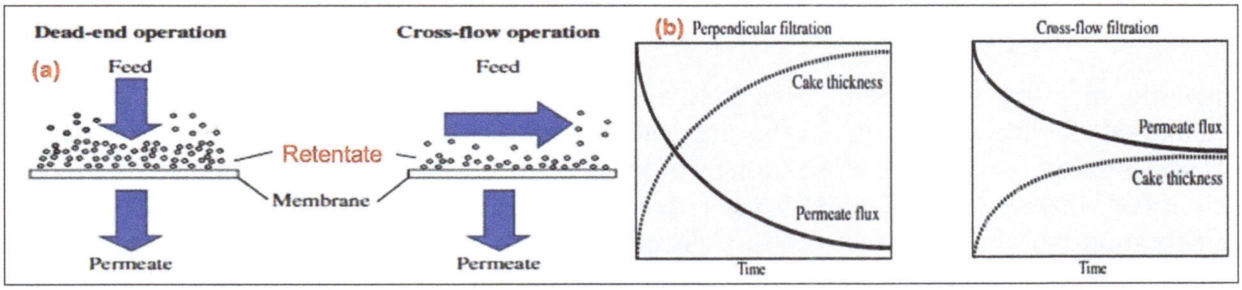

(a) Schematic of a dead-end filtration and cross-flow filtration, (b)Effect of perpendicular and cross-flow filtration on flux and cake thickness.

Membrane Generations

As per the technical development carried out till date, membrane materials can be classified in four generations.

- First generation: Cellulose acetate derived membranes, which were developed to desalinate seawater. They were susceptible to microorganisms and disinfectants as well as sensitive to (3-8) pH and (maximum 50 °C) temperature.
- Second generation: Developed from synthetic polymers like polysulfone or polyolefin derivatives), these are better than first generation as having better resistant towards hydrolysis, pH and temperatures, but have lower resistance to mechanical compacting.
- Third generation: Mineral membranes consisting ceramic material based on zirconium or alumina oxide are usually deposited on a graphite surface. These membranes have great mechanical strength, are chemically inert and can be operated on high pressures, wide pH range (0-14) and temperatures > 400 °C. Their main demerit of these membranes is their cost(expensive than other membrane materials), but the same is compensated by their long operational life.
- Fourth generation: Hybrid process i.e. combination of conventional Electrodialysis & different pore sizes membranes (E.D.+MF/UF/NF). The continuous electrophoresis with porous membranes (CEPM) is defined as an electrochemical process to separate charged organic molecules utilizing electric field as a driving force. Under the effect of an electric current, ions are transported from one solution to another through one or more semipermeable porous membranes.

Membrane Processing in Food Processing

Availability of few food materials in naturally pure form underlines the importance of different separation and purification processes in transforming food or food substances into safe and direct consumable form. For selected applications, membrane processing is superior to other conventional methods owing to its inherent advances such as low energy consumption, fewer and mild processing steps, greater separation efficiency, improved final product quality, eco-friendly and 'cleaner' processing i.e. better quality products at lowest cost with highest environment protection through minimum/low waste generation. Now a days, these processes have been become an integral part of different industries like functional food and nutraceuticals industry, agricultural, dairy and food and bio-product industries.

Membrane Application in Beer, Wine and Vinegar Production

Currently, membranes are being used in this area for clarification of beer, wine and vinegar as well as dealcoholization of beer. In the classical beer production process, as an alternate, reverse osmosis is used for beer dealcoholization prior to its clarification by microfiltration before pasteurization and bottling. Cross-flow MF is used to recover beer from the yeast-beer mixture (settled on fermentation tank bottom) and also to concentrate yeasts. The membrane system investment is usually compensated by the beer recovery. Moreover, MF now successfully employed for beer clarification i.e. for removal of yeasts, microorganisms and haze without adversely affecting its sensorial attributes.RO is used to tailor the alcohol content in beer to produce alcohol free or reduced alcohol (8-10 times less alcohol content) beer without affecting its flavour. It is carried out in four steps i.e. beer is pre-concentrated in permeate (water + alcohol) and retentate (concentrated beer + flavour) followed by diafiltration of the permeate and tailoring of alcohol content as per the desired taste using desalted and deoxygenized water; post treatment which includes the addition of hops and syrups in alcohol tailored beer to compensate the taste losses arises from the loss of taste carrier alcohols. These steps are performed ≤7 °C that results in the production of improved quality beer rather than conventional process.

Reverse osmosis is used for the must correction, rejuvenation and dealcoholization of wine, while MF/UF is used for the wine clarification during wine manufacture. Must composition is balanced with the help of RO, which increase sugar content in wine through concentration and at the same time enhances tannins and organoleptic components by 5-20%. It is an efficient method to maintain the quality of must obtained from the grapes harvested during rain (i.e. with moisture content) but the same is less effective in maintaining the quality of must made from the grapes of stalled maturity due to cold weather. The use of this particular method is depends on the regulation of various countries. During wine clarification, MF and UF provides an alternative to the classical fining substances, filters and also reduces the number of processing steps. Lipinzki reported that 0.20-0.50 μm and 0.45- 0.65 μm pore size MF membranes are usually used during the filtration of white and red wines, respectively. Rejuvenation of older wines is accomplished through RO process which removes the water, alcohol and other substances responsible for adverse aroma followed by the addition of mineral free water during diafiltration. Alcohol content of the wines is also tailored in the same manner as discussed above for beer.

Clarification of vinegar employing UF not only provides wider vinegar range with same sensorial attributes without turbidity, but also reduces number of operations and reduces storage time.

Membrane Application in Fruit Juice and Beverages Production

MF and UF are the economic and efficient alternatives to the classical filtration methods available for clarification of different juices. The use of membrane processing in beverage industry have several inherent advantages like improved product quality, reduced cost of production, better working environment, cleaner production with less waste generation and improved product safety. Cassano, reported several merits of membrane processes over classical beverages or fruit juice fining agents such as reduced energy consumption and cost, eradication of different filtration aids, mild thermal treatments that safeguard the product against thermal degradation, easier process and compact equipment design, fewer wastes, reduced chemicals requirement and enhanced productivity (96– 98%, juice recovery). Moreover, fruit juice clarification with MF and UF have substantial potential of cost saving. Annual production costs of a UF plant (capacity- 250 m^3 of juice/day) was estimated 79% lesser than a conventional plant. Different characteristics of conventional processes and MF filtration is compared and shown in Table.

Table: Clarification of fruit juices-comparison of traditional processes with microfiltration.

Processing method	Operating time	Fibres	Sensory and nutritional quality	Operating cost
Decanting	Very long	Poor	Poor	Intermediate
Dead-end filtration	Long	High	Intermediate	Intermediate
Centrifugation	Very short	Poor	High	High
Microfiltration	Short	High	High	Low

MF (pore sizes, 0.05–10 μm) is resembles classical coarse filtration, employed to filter juices in fibrous enriched pulp (in retentate) as well as microbes free clarified fraction (in permeate).

Moreover, it is known as cold – sterilization process. UF process is the alternate of the fining step used in classical method. It is wider membrane pores than MF yet retails the bacteria, fats, proteins and colloids but allow to pass the smaller molecules like vitamin, minerals and sugars. Prior to UF process, both pre and enzyme treatment of the juice is essential for better yield, high capacity and excellent product quality. Juices are first concentrated in RO, which decreases the moisture content to its half value but retains major amount of sugar (98– 99%) and volatile flavours (80-90%) in the retentate (having 20-25 ◦Brix) and same then concentrated > 75 ◦Brix in classical evaporation. This combination results in 60-75% saving of energy than traditional evaporation.

Membrane Applications in Functional Foods and Nutraceuticals

Global market of functional foods and nutraceuticals is rising rapidly. The future of recently evolved bio-activemolecules and nutraceutical ingredients is usually decided by the separation ability and efficiency with the retention of the desired characteristics/biological activity of these ingredients by the employed separation process with minimum inputs (i.e. energy + capital + labor cost). Separation as well as purification of such ingredients is highly expensive. Although, a number of challenges have to overcome to raise the concentration of a target substance in the end product from its raw material yet, several methods are accomplishing such tasks successfully. On commercial scale, membrane processing is now considered as low cost and effective tool to concentrate and purify several bioactive substances from different feed streams.

Table: Established markets and applications in nutraceutical and bioactive separation industries.

Process	System	Remarks
Ultrapure water, Water softening	RO/UF	Classical
Effluents	RO/NF/UF	Classical
Fine chemical processes	NF/UF	Developing rapidly
Milk/whey/milk bioactives	RO/NF/UF	Almost classical/Developing
Extract concentration	RO/NF	Restricted due to thecomplexity of processes
Emulsion separation	UF/MF	Classical but also expensive
Caustic recovery	Ceramic	Good technology for causticrecovery for all industries
Fruit juice clarification	MF	Established market
Oil/Water separation	UF/RO	Works well for non-emulsifiedmixtures
Alcohol purification	PV	Efficient inwater content <20% - developing, but good-potential
Enzyme recovery	UF/NF	Pharmaceutical industries –established but growing-market
Protein concentration and purification	UF/NF	Protein isolate fractions-Specialized applications
Waste water recovery	NF/RO	Commercial. Still developing
Wine clarification, Beer recovery	MF	Commercial
Wine sugar concentration	RO	Commercial
Color removal	NF	Commercial

Phospholipid removal fromcrude oils	NF/UF	Developing- Good Potential
Bio-active recovery from fruit and vegetable juices and herbs and botanicals	NF/RO	Good potential
Improve beverage stability	MF	Good application-eliminates heat treatment better flavor

Various membrane processes alone or in combination are used to produce numerous nutraceuticals and bioactive substances from lipids, carbohydrates and proteins protein based sources. Moreover, this technology offers many key advantages during processing of such ingredients which includes elimination of classical evaporation (that reduces their activity and highly energy consuming); reduces overall water needs by re-utilizing waste water; enhances profit by formation of new products and needs lesser floor space and investments. Table, represents the various application of membrane processing in different classical processes and in separation bio-actives.

Membrane Application in Vegetable Oil Processing

Traditional oil refining methods have several demerits like nutrient loss, needs higher amount of energy, water and chemicals; greater losses of neutral oil as well as the generation of more effluent. Vegetable oil refining employing membranes has been not only reported as an easier/simpler process but also the same offers different advantages like better nutrient retention, reduced demand of energy and added chemicals, milder process parameters and cleaner production. Several researchers have been used membrane processes for the recovery of solvents from the micelle, degumming, bleaching, deacidification as well as hydrolysis of fats and oils.

Membrane Application in Dairy Processing

Dairy industry is one of the early adaptor of the membrane processes. It has greatly benefited by membranes as a number of applications like removal of water/concentration, liquid-liquid as well as solid-liquid separations are routinely performed. From 1960, the major milestones in membrane development and their application in dairy processing is presented in table; different filtration spectrums available for the separation of milk constituents is shown in table.

Table: Milestones in the development of membrane technology and its applications in dairy processes since 1960s.

	Advances in membrane technology	Applications of membranes in dairy processing
1960s	Development of reproducible membranes by manufacturers.	–
	Materials with improved chemical resistance (from CA to PS) First design of sanitary modules.	Design of whey pre-treatments to prevent membrane fouling Development of processes for the UF of acid whey Development of the first UF-based cheese manufacture processes.
1980s	Improvement of membrane system hardware (module designs, spacers, anti-telescoping devices). Development of commercial inorganic (ceramic) membranes.	Using UF or RO membranes to concentrate milk on farm. Defatting of whey (WPI manufacture, recovery of minor compounds). Separation of β-lactoglobulin & αlactalbumin Desalting whey with loose-RO (NF) membranes.

1990s	Improvement of hydrodynamics of MF membranes (UTP). Porosity gradient membranes. Control of particle's deposition (vibration, rotating disk, Dean's vortices, static mixer) Functionalized membranes (ion exchange).	Removing spores from cheese milk and whey. Defatting whey. Separating casein micelles from milk (ideal whey). Extending milk's shelf life (ESL milk). Fractionating hydrolysates using UF/NF membranes.

Table: Filtration spectrum available for the separation of milk constituents.

Pore size	Separation mechanism	Operating pressure (MPa)	Membrane materials	Module configuration	Separation domain	Membrane-based commercial dairy ingredient
MF >0.1 μm	Sieving	0.01-0.2	Inorganic Polymeric	T, MC	Somatic cells, bacteria, spores Fat globules Casein micelles	Micellar casein, Native whey proteins
UF 1-500nm	Sieving & charge	0.1-1		T, HF, SW, PF	Soluble proteins Caseinomacropeptide	WPC, WPI, MPC, β-Lg, α-La
NF0.1- 1nm	Sieving & charge	1.5-3		T, HF, SW, PF	Indigenous peptides Salts (divalentcations)	Bioactive Milk & whey proteins hydrolysates, Glycomacropeptide
RO<0.1 nm	Sieving & Diffusion	3-5		SW, PF	Salts, Water removal	Lactose, Concentration of Whey permeate Delactosed, deproteinized whey
Polymeric: cellulosic, polysulfone, polyamide; inorganic: ceramic, carbon-supported zirconium oxide, stainless; T, tubular; MC, multichannel; HF, hollow fiber; SW, spiral wound; PF, plate and frame ;WPC, whey protein concentrate; WPI, whey protein isolate; MPC, milk protein concentrate.						

The key membrane applications in dairy industry is categorized in three main classes which are shown in Figure. Moreover, the use of different classical membrane processes particularly in milk and whey processing is shown in Figures, respectively.

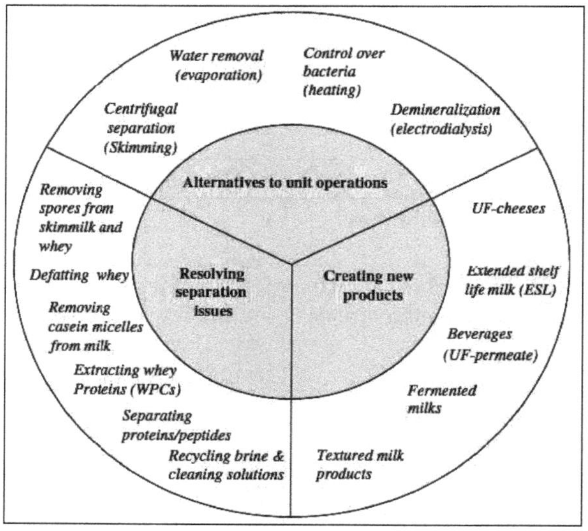

Membrane processes in the dairy industry: a look at the applications.

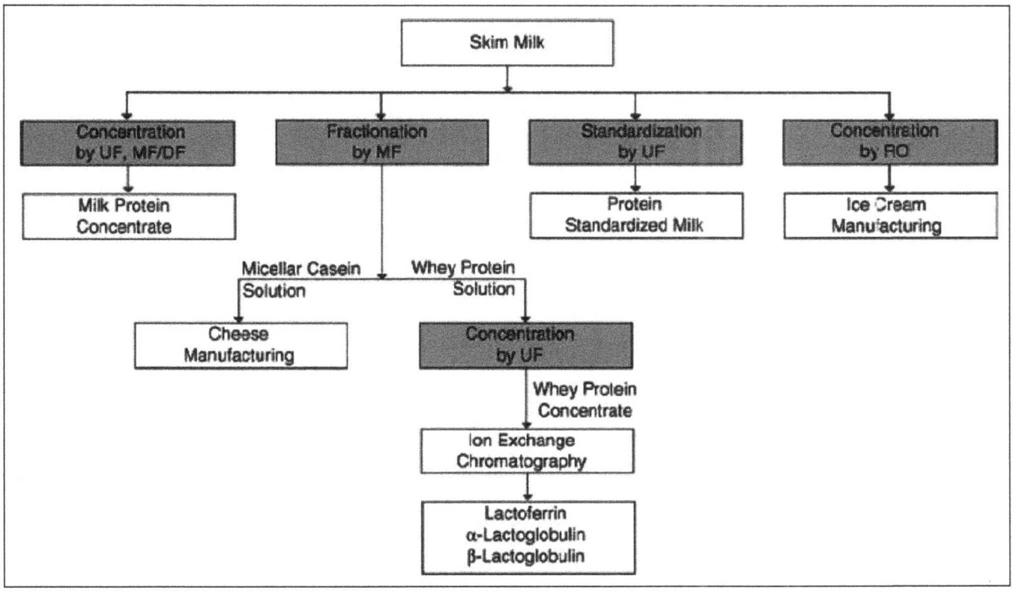

Applications of membrane technology in milk processing.

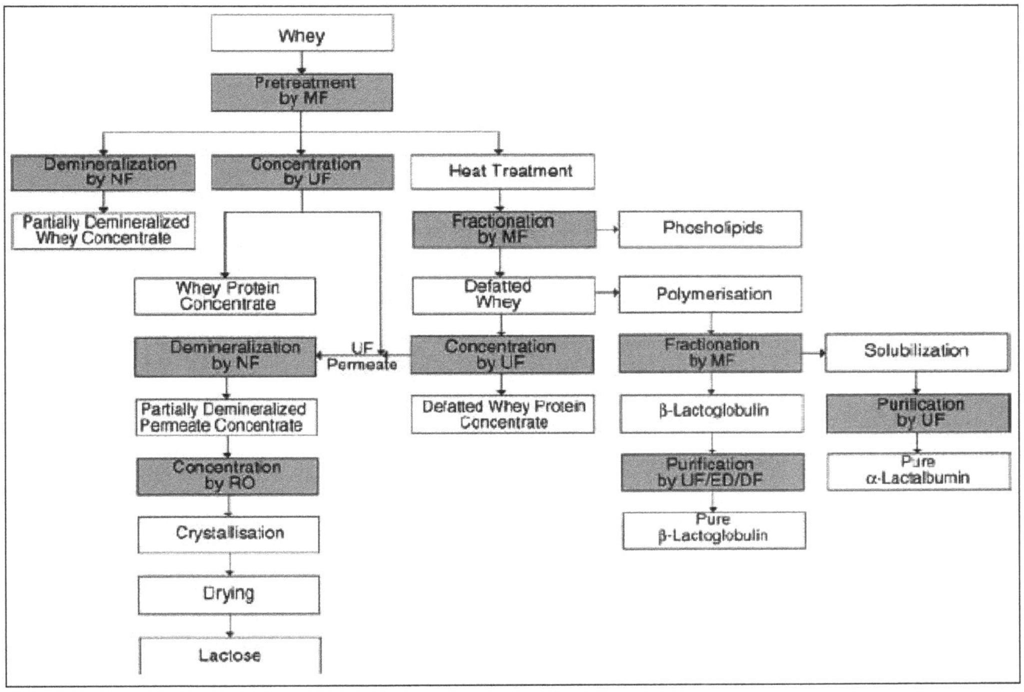

Applications of membrane technology in whey processing.

Air Classification Techniques

Air classification is a technological method of separating particulates according to their size, as represented by particle equivalent diameter. The latter may be either geometric or hydrodynamic/ aerodynamic equivalent diameter, depending on the size-measurement method used. The initial or original (o) material is divided into two products: coarse (c) and fine (f). Separation is characterized by a cut size d_{50}, which is a border between these fractions. Particles with sizes above d_{50} constitute coarse fraction and those with sizes below d_{50} are fines. Due to various stochastic factors (air turbulence, particle collisions), some fines get into the coarse product and vice versa. Quality of particle separation is characterized by mass contents of each fraction in the suitable product. An interior space of a classifying device, where solid particles interact with air stream, is called separation zone. Four basic separation zones are there, which includes gravitational-counter flow, gravitational-cross flow, centrifugal-counter flow and centrifugal-cross flow zone.

Gravitational-Counter Flow Zone

The gravitational-counter flow zone exists in air stream rising up with velocity u inside a vertical chamber with parallel walls. Particles experience gravity G and drag Fd forces acting in opposite directions. Coarse particles, having terminal settling velocity VTS larger than air flow velocity, move downwards, against the air stream, and fines rise with the stream. The cut size d50 is calculated from the equilibrium G = Fd combined with zero particle's absolute velocity, i.e. VTS = u. Theoretically, the cut size particles are indefinitely suspended in the separation zone. Practically all particles are separated due to unavoidable stochastic factors.

Gravitational-Cross Flow Zone

A horizontal air stream entering this zone expands within a chamber and converges towards the outlet. Separation occurs due do particles' vertical motion across the air stream. Each particle falls at its own terminal settling velocity. Accordingly, due to the horizontal motion they land on the bottom at different chamber locations: coarse-close to the inlet, smaller-close to the outlet. Coarse fraction is gathered on the bottom of the chamber fines do not reach the bottom and are carried away by the air flow through the outlet. Thus, in the gravitational-cross flow zone particle separation is a two-dimensional process, where their trajectories are in fact ballistic tracks. Unlike the counter flow zone, the cut size depends not only on the particle's terminal velocity, but mainly on the chamber length and height. These parameters are chosen in such a way that particles with the cut size land at the farthest point A of the bottom.

Centrifugal-Counter Flow Zone

This separation zone is characterized by a flat air vortex prevailing in a cylindrical chamber with tangential inlet and central outlet. Particles fed tangentially with air stream are involved in rotation. Due to the centrifugal (inertial) force Fc their trajectories gradually deviate from the air streamlines. Separation is governed by the balance between Fc and the drag force component Fdr induced by the radial air movement. Coarse particles drift towards the chamber walls, where they fall down to the outlet. Fines move towards the chamber axis where they are eventually carried out by the air stream. Particles with cut size theoretically move with zero radial velocity, but practically go to one of the products. As the centrifugal force depends on the radial position, so does the cut size d50. This distinguishes the separation process in the centrifugal-counter flow zone from the separation occurring in the gravitational-counter flow zone. To keep the cut size constant along the radial direction, the classifier should be designed to provide constant tangential air velocity. This feature is characteristic of the Archimedean spiral or free vortex flow.

Centrifugal-Cross Flow Zone

This zone may be created in an air vortex for example in a cylindrical chamber with inlet and outlet placed on the opposite sides of the chamber axis. The fed particles enter with the air stream through whirl blades, which create a screw-type flow. The blade profiles should be designed such that the radial component of the air velocity everywhere vanishes. Particles rotate while drifting radially towards the chamber walls with size-dependent velocities. As they simultaneously proceed along the chamber axis they reach the wall in different locations. Larger particles arrive to the wall faster; smaller ones make it farther along the axial direction. The chamber is designed in such a way that the cut size particles reach the periphery at its edge (point A). The separation process in the centrifugal-cross flow zone is similar to that in cyclones; so these devices have similar design principles.

No universal design method can be recommended for separation of different materials, as each device type has its specific technical requirements and suitable for specific product. The separation zones described above serve as a basis for systematizing air classifiers, which may have several zones used in combination in the same device. Air classifiers may have external or internal aerodynamic cycles. The former devices are constructed independently of dust collectors and fans and connected to them via pipelines. The internal-cycle (recirculating-air) devices contain these elements in a single unit, and hence need less production area and normally require lower investments. Air classifiers may also be distinguished by method of powder feed. Particles may be fed either within air stream, or separately by freely pouring them within the device. Other systematizing may be done with respect to the air stream form, classifier design features and operation mode (continuous or batch). Below we review principal schemes of most commonly used air classifiers to summarize the most important features of their design and operation. Modern trends of classifier design account for stochastic nature of separation and aerodynamic parameters of equipment, creation of precise air velocity field, repeated cleaning of both fractions, arrangement of several classifying stages, etc.

Classification

- Gravitational air classifiers:
    - Counter flow classifiers with vertical (rising) flows.
    - Crossflow classifiers with horizontal air stream.
- Cascade air classifiers.
- Fluidized bed classifiers.
- Inertial air classifiers.
- Centrifugal air classifiers:
    - Vortex air classifiers.
    - Gravitational–centrifugal air classifiers.
    - Rotor classifiers.
    - Circulating air classifiers.

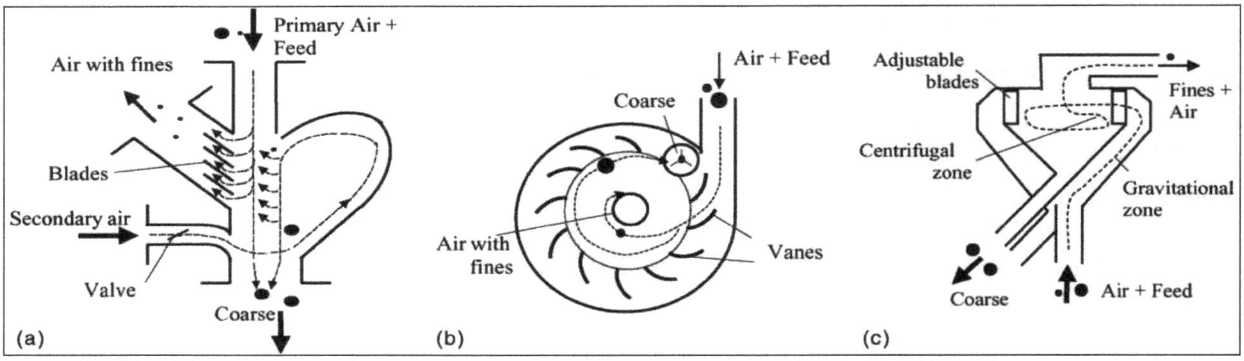

(a) Franken/Buell gravitational-inertial classifier, (b) Vortex classifier,
(c) Gravitational–centrifugal classifier.

Counter Flow Classifiers with Vertical (Rising) Flows

These classifiers are used only for separation of coarse bulk materials, containing two types of particles of well differing sizes, e.g. coarse sand or pebbles of 1–10mm mixed with dust 0.1–0.2 mm.

Cross Flow Classifiers with Horizontal Air Stream

Separation in these devices known as winnowing machines is achieved due to the difference between trajectories of coarse and fine particles in the separation zone. Their operation and efficiency are strongly affected by the stochastic factors of the process, in particular by uncertainties in feeding and particles' aerodynamic interactions. In most cases coarse particles prevent proper classification of fines. Separation efficiency of these devices is usually low. They are normally used for separation of solid particles according to densities (e.g. grain from peel), rather than by size. Sometimes cross flow separation in horizontal streams is used in combination with other separation principles.

Cascade Air Classifiers

These devices, known as Zigzag classifiers, are cascades of several inclined branch pipes of a rectangular cross section arranged to form a vertical zigzag channel. Air flowing from below scavenges the particles fed from above in cross-current direction. The flow field within the chamber has vortexes in the turn points, where the fines are dragged by the stream, and the coarse particles cross it towards the opposite wall and pour down. Separation occurs in a crossflow process. Sequential pipes' arrangement enables multiple scavenging to improve the separation efficiency. Several types of Zigzag classifier are produced for separation of ferruginous ore, fertilizers, plastic, wooden dust, etc. at cut sizes ranging from 0.1 to 10 mm. The number of parallel chambers varies from one to 16 with output ranging from 1 to10 t/h. Although multiple scavenging improves separation, coarse particles crossing the air stream on each cascade step collide and mix with fines. This precludes sharp separation and decreases the efficiency. Another disadvantage of the device is its high aerodynamic resistance (pressure drop of about 2.5–3.5 kPa), resulting from the vortexes formed in cascades' turn points. Cascade principle was also combined with cross-count horizontal scavenging in a vertical vessel divided into separation and precipitation chambers, separated by a permeable grid with concave shelves.

Fluidized Bed Classifiers

One disadvantage of the fluidized bed classifiers with cylindrical chambers of large diameters (above 1m) is formation of dead zones within the fluidized bed far from the axis, where the particles do not move. Fluidized bed classification plant constructed and it has 30–200 kg/h feeder output, air flow rate of up to 400m3/h allowing to change the cut from 50 to1000μm. The efficiency is about 90% and above, depending on the material. The grid of changeable slope allows controlling the rate of discharge of coarse particles out of the separator. Investigations showed good performance stability of the classifier in the wide range (up to four times) of feed rate. These devices have both products' recoveries higher at least by 15–20% with respect to other gravitational classifiers.

Inertial Air Classifiers

These devices are employed for fine powders (40–400 μm) and also coarser particles. In the Franken inertial separators particle-air mixture enters from the top and falls down in the inlet pipe. The air turns abruptly to the outlet pipe inclined by about 45–50° and carries the fine particles. Coarse fraction proceeds straight down because of high inertia towards the discharge pipe. On its way this material undergoes repeated separation in a secondary air flow carrying fines into a curvilinear chamber. Then they exit via the outlet pipe with blades mounted for better air distribution.

Centrifugal Air Classifiers

Two main types of these classifiers are known, distinguished by method of creating air rotation. In vortex classifiers the air is rotated by stationary vanes (or tangential pipes); in rotor classifiers this is done by rotating blades. The latter devices may have open or closed aerodynamic cycle. Choice of the classifier type is dictated by product requirements, technological and economic factors. Rational design and improving separation quality is done by creating a well-defined and stable air flow, reducing turbulence, eliminating particles collisions, controlling feed and repeated cleaning of both fractions. Among gravitational type classifiers, fluidized-bed based devices provide stable operation and sharp separation in the cut size range of 50–1000 μm. Smaller-size particles may be separated by rotor classifiers, enabling easy cut-size control and providing products' recoveries up to 80%.

References

- Advances-in-Food-Science-and-Technology, biosensors: inflibnet.ac.in, Retrieved 02, January 2020
- Adulteration-the-underbelly-of-food-processing: quacktrack.org, Retrieved 23, July 2020

Chapter 5

Genetically Modified Food

The foods which are sourced from organisms whose DNA has been changed through genetic engineering are known as genetically modified foods. Genetic engineering techniques enable the production of crops which are resistant to pathogens and herbicides as well as possess better nutrition profiles. This chapter closely examines the key concepts related to the advantages and disadvantages of genetically modified crops to provide an extensive understanding of the subject.

Genetically modified foods or GM foods, also genetically engineered foods, are foods produced from organisms that have had changes introduced into their DNA using the methods of genetic engineering. Genetic engineering techniques allow for the introduction of new traits as well as greater control over traits than previous methods such as selective breeding and mutation breeding.

Commercial sale of genetically modified foods began in 1994, when Calgene first marketed its unsuccessful Flavr Savr delayed-ripening tomato. Most food modifications have primarily focused on cash crops in high demand by farmers such as soybean, corn, canola, and cotton. Genetically modified crops have been engineered for resistance to pathogens and herbicides and for better nutrient profiles. GM livestock have been developed, although as of November 2013 none were on the market.

There is a scientific consensus that currently available food derived from GM crops poses no greater risk to human health than conventional food, but that each GM food needs to be tested on a case-by-case basis before introduction. Nonetheless, members of the public are much less likely than scientists to perceive GM foods as safe. The legal and regulatory status of GM foods varies by country, with some nations banning or restricting them, and others permitting them with widely differing degrees of regulation.

However, there are ongoing public concerns related to food safety, regulation, labelling, environmental impact, research methods, and the fact that some GM seeds are subject to intellectual property rights owned by corporations.

Definition

Genetically modified foods, GM foods or genetically engineered foods, are foods produced from organisms that have had changes introduced into their DNA using the methods of genetic engineering as opposed to traditional cross breeding. In the US, the Department of Agriculture (USDA) and the Food and Drug Administration (FDA) favor the use of "genetic engineering" over "genetic modification" as the more precise term; the USDA defines genetic modification to include "genetic engineering or other more traditional methods."

According to the World Health Organization, "Genetically modified organisms (GMOs) can be defined as organisms (i.e. plants, animals or microorganisms) in which the genetic material (DNA) has been altered in a way that does not occur naturally by mating and/or natural recombination. The technology is often called 'modern biotechnology' or 'gene technology', sometimes also 'recombinant DNA technology' or 'genetic engineering'. ... Foods produced from or using GM organisms are often referred to as GM foods."

History

Human-directed genetic manipulation of food began with the domestication of plants and animals through artificial selection at about 10,500 to 10,100 BC. The process of selective breeding, in which organisms with desired traits (and thus with the desired genes) are used to breed the next generation and organisms lacking the trait are not bred, is a precursor to the modern concept of genetic modification (GM). With the discovery of DNA in the early 1900s and various advancements in genetic techniques through the 1970s it became possible to directly alter the DNA and genes within food.

Genetically modified microbial enzymes were the first application of genetically modified organisms in food production and were approved in 1988 by the US Food and Drug Administration. In the early 1990s, recombinant chymosin was approved for use in several countries. Cheese had typically been made using the enzyme complex rennet that had been extracted from cows' stomach lining. Scientists modified bacteria to produce chymosin, which was also able to clot milk, resulting in cheese curds.

The first genetically modified food approved for release was the Flavr Savr tomato in 1994. Developed by Calgene, it was engineered to have a longer shelf life by inserting an antisense gene that delayed ripening. In 1995, *Bacillus thuringiensis* (Bt) Potato was approved for cultivation, making it the first pesticide producing crop to be approved in the USA. Other genetically modified crops receiving marketing approval in 1995 were: canola with modified oil composition, Bt maize, cotton resistant to the herbicide bromoxynil, Bt cotton, glyphosate-tolerant soybeans, virus-resistant squash, and another delayed ripening tomato.

With the creation of golden rice in 2000, scientists had genetically modified food to increase its nutrient value for the first time.

By 2010, 29 countries had planted commercialized biotech crops and a further 31 countries had granted regulatory approval for transgenic crops to be imported. The US was the leading country in the production of GM foods in 2011, with twenty-five GM crops having received regulatory approval. In 2015, 92% of corn, 94% of soybeans, and 94% of cotton produced in the US were genetically modified strains.

The first genetically modified animal to be approved for food use was AquAdvantage salmon in 2015. The salmon were transformed with a growth hormone-regulating gene from a Pacific Chinook salmon and a promoter from an ocean pout enabling it to grow year-round instead of only during spring and summer.

In April 2016, a white button mushroom (*Agaricus bisporus*) modified using the CRISPR technique received *de facto* approval in the United States, after the USDA said it would not have to go through the agency's regulatory process. The agency considers the mushroom exempt because the editing process did not involve the introduction of foreign DNA.

The most widely planted GMOs are designed to tolerate herbicides. By 2006 some weed populations had evolved to tolerate some of the same herbicides. Palmer amaranth is a weed that competes with cotton. A native of the southwestern US, it traveled east and was first found resistant to glyphosate in 2006, less than 10 years after GM cotton was introduced.

Process

Genetically engineered organisms are generated and tested in the laboratory for desired qualities. The most common modification is to add one or more genes to an organism's genome. Less commonly, genes are removed or their expression is increased or silenced or the number of copies of a gene is increased or decreased.

Once satisfactory strains are produced, the producer applies for regulatory approval to field-test them, called a "field release." Field-testing involves cultivating the plants on farm fields or growing animals in a controlled environment. If these field tests are successful, the producer applies for regulatory approval to grow and market the crop. Once approved, specimens (seeds, cuttings, breeding pairs, etc.) are cultivated and sold to farmers. The farmers cultivate and market the new strain. In some cases, the approval covers marketing but not cultivation.

According to the USDA, the number of field releases for genetically engineered organisms has grown from four in 1985 to an average of about 800 per year. Cumulatively, more than 17,000 releases had been approved through September 2013.

Crops

Fruits and Vegetables

3 views of the Sunset papaya cultivar, which was genetically modified to create the SunUp cultivar, resistant to PRSV.

Papaya was genetically modified to resist the ringspot virus. 'SunUp' is a transgenic red-fleshed Sunset papaya cultivar that is homozygous for the coat protein gene PRSV; 'Rainbow' is a yellow-fleshed F1 hybrid developed by crossing 'SunUp' and nontransgenic yellow-fleshed 'Kapoho'. The New York Times stated, "in the early 1990s, Hawaii's papaya industry was facing disaster because of the deadly papaya ringspot virus. Its single-handed savior was a breed engineered to be resistant to the virus. Without it, the state's papaya industry would have collapsed. Today, 80% of Hawaiian papaya is genetically engineered, and there is still no conventional or organic method to control ringspot virus." The GM cultivar was approved in 1998. In China, a transgenic PRSV-resistant papaya was developed by South China Agricultural University and was first approved for commercial planting in 2006; as of 2012 95% of the papaya grown in Guangdong province and 40% of the papaya grown in Hainan province was genetically modified.

The New Leaf potato, brought to market by Monsanto in the late 1990s, was developed for the fast food market. It was withdrawn in 2001 after retailers rejected it and food processors ran into export problems.

As of 2005, about 13% of the Zucchini (a form of squash) grown in the US was genetically modified to resist three viruses; that strain is also grown in Canada.

Plums genetically engineered for resistance to plum pox, a disease carried by aphids.

In 2011, BASF requested the European Food Safety Authority's approval for cultivation and marketing of its Fortuna potato as feed and food. The potato was made resistant to late blight by adding resistant genes blb1 and blb2 that originate from the Mexican wild potato Solanum bulbocastanum. In February 2013, BASF withdrew its application.

In 2013, the USDA approved the import of a GM pineapple that is pink in color and that "overexpresses" a gene derived from tangerines and suppress other genes, increasing production of lycopene. The plant's flowering cycle was changed to provide for more uniform growth and quality. The fruit "does not have the ability to propagate and persist in the environment once they have been harvested," according to USDA APHIS. According to Del Monte's submission, the pineapples are commercially grown in a "monoculture" that prevents seed production, as the plant's flowers aren't exposed to compatible pollen sources. Importation into Hawaii is banned for "plant sanitation" reasons.

In 2014, the USDA approved a genetically modified potato developed by J.R. Simplot Company that contained ten genetic modifications that prevent bruising and produce less acrylamide when fried. The modifications eliminate specific proteins from the potatoes, via RNA interference, rather than introducing novel proteins.

In February 2015 Arctic Apples were approved by the USDA, becoming the first genetically modified apple approved for sale in the US. Gene silencing is used to reduce the expression of polyphenol oxidase (PPO), thus preventing the fruit from browning.

Corn

Corn used for food and ethanol has been genetically modified to tolerate various herbicides and to express a protein from Bacillus thuringiensis (Bt) that kills certain insects. About 90% of the corn grown in the U.S. was genetically modified in 2010. In the US in 2015, 81% of corn acreage contained the Bt trait and 89% of corn acreage contained the glyphosate-tolerant trait. Corn can be processed into grits, meal and flour as an ingredient in pancakes, muffins, doughnuts, breadings and batters, as well as baby foods, meat products, cereals and some fermented products. Corn-based masa flour and masa dough are used in the production of taco shells, corn chips and tortillas.

Soy

Genetically modified soybean has been modified to tolerate herbicides and produce healthier oils. In 2015, 94% of soybean acreage in the U.S. was genetically modified to be glyphosate-tolerant.

Derivative Products

Corn Starch and starch Sugars, Including Syrups

Starch or amylum is a polysaccharide produced by all green plants as an energy store. Pure starch is a white, tasteless and odourless powder. It consists of two types of molecules: the linear and helical amylose and the branched amylopectin. Depending on the plant, starch generally contains 20 to 25% amylose and 75 to 80% amylopectin by weight.

Starch can be further modified to create modified starch for specific purposes, including creation of many of the sugars in processed foods. They include:

- Maltodextrin, a lightly hydrolyzed starch product used as a bland-tasting filler and thickener.
- Various glucose syrups, also called corn syrups in the US, viscous solutions used as sweeteners and thickeners in many kinds of processed foods.
- Dextrose, commercial glucose, prepared by the complete hydrolysis of starch.
- High fructose syrup, made by treating dextrose solutions with the enzyme glucose isomerase, until a substantial fraction of the glucose has been converted to fructose.
- Sugar alcohols, such as maltitol, erythritol, sorbitol, mannitol and hydrogenated starch hydrolysate, are sweeteners made by reducing sugars.

Lecithin

Lecithin is a naturally occurring lipid. It can be found in egg yolks and oil-producing plants. it is an emulsifier and thus is used in many foods. Corn, soy and safflower oil are sources of lecithin, though the majority of lecithin commercially available is derived from soy.Sufficiently processed lecithin is often undetectable with standard testing practices.According to the FDA, no evidence shows or suggests hazard to the public when lecithin is used at common levels. Lecithin added to foods amounts to only 2 to 10 percent of the 1 to 5 g of phosphoglycerides consumed daily on average. Nonetheless, consumer concerns about GM food extend to such products.This concern led to policy and regulatory changes in Europe in 2000, when Regulation (EC) 50/2000 was passed which required labelling of food containing additives derived from GMOs, including lecithin. Because of the difficulty of detecting the origin of derivatives like lecithin with current testing practices, European regulations require those who wish to sell lecithin in Europe to employ a comprehensive system of Identity preservation (IP).

Sugar

The US imports 10% of its sugar, while the remaining 90% is extracted from sugar beet and sugarcane. After deregulation in 2005, glyphosate-resistant sugar beet was extensively adopted in the United States. 95% of beet acres in the US were planted with glyphosate-resistant seed in 2011. GM sugar beets are approved for cultivation in the US, Canada and Japan; the vast majority are grown in the US. GM beets are approved for import and consumption in Australia, Canada, Colombia, EU, Japan, Korea, Mexico, New Zealand, Philippines, Russian Federation and Singapore. Pulp from the refining process is used as animal feed. The sugar produced from GM sugarbeets contains no DNA or protein—it is just sucrose that is chemically indistinguishable from sugar produced from non-GM sugarbeets. Independent analyses conducted by internationally recognized laboratories found that sugar from Roundup Ready sugar beets is identical to the sugar from comparably grown conventional (non-Roundup Ready) sugar beets. And, like all sugar, sugar from Roundup Ready sugar beets contains no genetic material or detectable protein (including the protein that provides glyphosate tolerance).

Vegetable Oil

Most vegetable oil used in the US is produced from GM crops canola, corn, cotton and soybeans. Vegetable oil is sold directly to consumers as cooking oil, shortening and margarine and is used in prepared foods. There is a vanishingly small amount of protein or DNA from the original crop in vegetable oil. Vegetable oil is made of triglycerides extracted from plants or seeds and then refined and may be further processed via hydrogenation to turn liquid oils into solids. The refining process removes all, or nearly all non-triglyceride ingredients.

Other Uses

Animal feed

Livestock and poultry are raised on animal feed, much of which is composed of the leftovers from processing crops, including GM crops. For example, approximately 43% of a canola seed is oil. What remains after oil extraction is a meal that becomes an ingredient in animal feed and contains canola protein. Likewise, the bulk of the soybean crop is grown for oil and meal. The high-protein defatted and toasted soy meal becomes livestock feed and dog food. 98% of the US soybean crop goes for livestock feed. In 2011, 49% of the US maize harvest was used for livestock feed (including the percentage of waste from distillers grains). "Despite methods that are becoming more and more sensitive, tests have not yet been able to establish a difference in the meat, milk, or eggs of animals depending on the type of feed they are fed. It is impossible to tell if an animal was fed GM soy just by looking at the resulting meat, dairy, or egg products. The only way to verify the presence of GMOs in animal feed is to analyze the origin of the feed itself."

A 2012 literature review of studies evaluating the effect of GM feed on the health of animals did not find evidence that animals were adversely affected, although small biological differences were occasionally found. The studies included in the review ranged from 90 days to two years, with several of the longer studies considering reproductive and intergenerational effects.

Proteins

Rennet is a mixture of enzymes used to coagulate milk into cheese. Originally it was available only from the fourth stomach of calves, and was scarce and expensive, or was available from microbial sources, which often produced unpleasant tastes. Genetic engineering made it possible to extract rennet-producing genes from animal stomachs and insert them into bacteria, fungi or yeasts to make them produce chymosin, the key enzyme. The modified microorganism is killed after fermentation. Chymosin is isolated from the fermentation broth, so that the Fermentation-Produced Chymosin (FPC) used by cheese producers has an amino acid sequence that is identical to bovine rennet. The majority of the applied chymosin is retained in the whey. Trace quantities of chymosin may remain in cheese.

FPC was the first artificially produced enzyme to be approved by the US Food and Drug Administration. FPC products have been on the market since 1990 and as of 2015 had yet to be surpassed in commercial markets. In 1999, about 60% of US hard cheese was made with FPC. Its global market share approached 80%. By 2008, approximately 80% to 90% of commercially made cheeses in the US and Britain were made using FPC. The most widely used FPC is produced either by the fungus *Aspergillus niger* (CHY-MAX®).

In some countries, recombinant (GM) bovine somatotropin (also called rBST, or bovine growth hormone or BGH) is approved for administration to increase milk production. rBST may be present in milk from rBST treated cows, but it is destroyed in the digestive system and even if directly injected into the human bloodstream, has no observable effect on humans. The FDA, World Health Organization, American Medical Association, American Dietetic Association and the National Institutes of Health have independently stated that dairy products and meat from rBST-treated cows are safe for human consumption. However, on 30 September 2010, the United States Court of Appeals, Sixth Circuit, analyzing submitted evidence, found a "compositional difference" between milk from rBGH-treated cows and milk from untreated cows. The court stated that milk from rBGH-treated cows has: increased levels of the hormone Insulin-like growth factor 1 (IGF-1); higher fat content and lower protein content when produced at certain points in the cow's lactation cycle; and more somatic cell counts, which may "make the milk turn sour more quickly."

Livestock

Genetically modified livestock are organisms from the group of cattle, sheep, pigs, goats, birds, horses and fish kept for human consumption, whose genetic material (DNA) has been altered using genetic engineering techniques. In some cases, the aim is to introduce a new trait to the animals which does not occur naturally in the species, i.e. transgenesis.

A 2003 review published on behalf of Food Standards Australia New Zealand examined transgenic experimentation on terrestrial livestock species as well as aquatic species such as fish and shellfish. The review examined the molecular techniques used for experimentation as well as techniques for tracing the transgenes in animals and products as well as issues regarding transgene stability.

Some mammals typically used for food production have been modified to produce non-food products, a practice sometimes called Pharming.

Salmon

A GM salmon, awaiting regulatory approval since 1997, was approved for human consumption by the American FDA in November 2015, to be raised in specific land-based hatcheries in Canada and Panama.

Recombinant Food-Grade Organisms for Healthcare

The use of genetically modified food-grade organisms as recombinant vaccine expression hosts and delivery vehicles can open new avenues for vaccinology. Considering that oral immunization is a beneficial approach in terms of costs, patient comfort, and protection of mucosal tissues, the use of food-grade organisms can lead to highly advantageous vaccines in terms of costs, easy administration, and safety. The organisms currently used for this purpose are bacteria (Lactobacillus and Bacillus), yeasts, algae, plants, and insect species. Several such organisms are under clinical evaluation, and the current adoption of this technology by the industry indicates a potential to benefit global healthcare systems.

Health and Safety

There is a scientific consensus that currently available food derived from GM crops poses no greater risk to human health than conventional food, but that each GM food needs to be tested on a case-by-case basis before introduction. Nonetheless, members of the public are much less likely than scientists to perceive GM foods as safe.

Opponents claim that long-term health risks have not been adequately assessed and propose various combinations of additional testing, labeling or removal from the market. The advocacy group European Network of Scientists for Social and Environmental Responsibility (ENSSER), disputes the claim that "science" supports the safety of current GM foods, proposing that each GM food must be judged on case-by-case basis. The Canadian Association of Physicians for the Environment called for removing GM foods from the market pending long term health studies. Multiple disputed studies have claimed health effects relating to GM foods or to the pesticides used with them.

Testing

The legal and regulatory status of GM foods varies by country, with some nations banning or restricting them, and others permitting them with widely differing degrees of regulation. Countries such as the United States, Canada, Lebanon and Egypt use *substantial equivalence* to determine if further testing is required, while many countries such as those in the European Union, Brazil and China only authorize GMO cultivation on a case-by-case basis. In the U.S. the FDA determined that GMO's are "Generally Recognized as Safe" (GRAS) and therefore do not require additional testing if the GMO product is substantially equivalent to the non-modified product. If new substances are found, further testing may be required to satisfy concerns over potential toxicity, allergenicity, possible gene transfer to humans or genetic outcrossing to other organisms.

Regulation

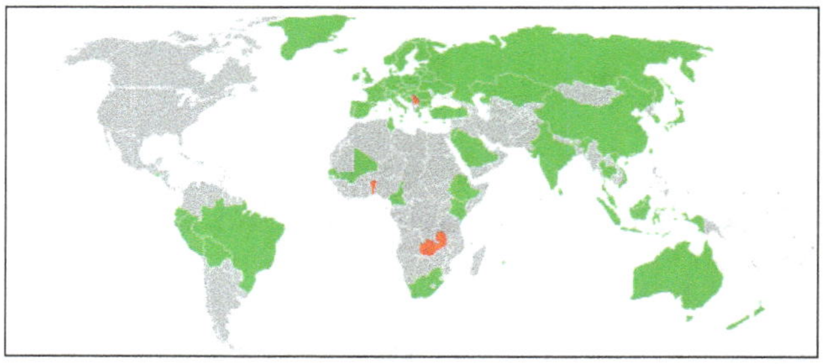

Green: Mandatory labeling required; Red:Ban on import and cultivation of genetically engineered food.

Government regulation of GMO development and release varies widely between countries. Marked differences separate GMO regulation in the U.S. and GMO regulation in the European Union. Regulation also varies depending on the intended product's use. For example, a crop not intended for food use is generally not reviewed by authorities responsible for food safety.

United States Regulations

In the U.S., three government organizations regulate GMOs. The FDA checks the chemical composition of organisms for potential allergens. The United States Department of Agriculture (USDA) supervises field testing and monitors the distribution of GM seeds. The United States Environmental Protection Agency (EPA) is responsible for monitoring pesticide usage, including plants modified to contain proteins toxic to insects. Like USDA, EPA also oversees field testing and the distribution of crops that have had contact with pesticides to ensure environmental safety. In 2015 the Obama administration announced that it would update the way the government regulated GM crops.

In 1992 FDA published "Statement of Policy: Foods derived from New Plant Varieties." This statement is a clarification of FDA's interpretation of the Food, Drug, and Cosmetic Act with respect to foods produced from new plant varieties developed using recombinant deoxyribonucleic acid (rDNA) technology. FDA encouraged developers to consult with the FDA regarding any bioengineered foods in development. The FDA says developers routinely do reach out for consultations. In 1996 FDA updated consultation procedures.

Labeling

As of 2015, 64 countries require labeling of GMO products in the marketplace.

US and Canadian national policy is to require a label only given significant composition differences or documented health impacts, although some individual US states (Vermont, Connecticut and Maine) enacted laws requiring them.

In some jurisdictions, the labeling requirement depends on the relative quantity of GMO in the product. A study that investigated voluntary labeling in South Africa found that 31% of products labeled as GMO-free had a GM content above 1.0%.

In Europe all food (including processed food) or feed that contains greater than 0.9% GMOs must be labelled.

Detection

Testing on GMOs in food and feed is routinely done using molecular techniques such as PCR and bioinformatics.

In a January 2010 paper, the extraction and detection of DNA along a complete industrial soybean oil processing chain was described to monitor the presence of Roundup Ready (RR) soybean: "The amplification of soybean lectin gene by end-point polymerase chain reaction (PCR) was successfully achieved in all the steps of extraction and refining processes, until the fully refined soybean oil. The amplification of RR soybean by PCR assays using event-specific primers was also achieved for all the extraction and refining steps, except for the intermediate steps of refining (neutralisation, washing and bleaching) possibly due to sample instability. The real-time PCR assays using specific probes confirmed all the results and proved that it is possible to detect and quantify genetically modified organisms in the fully refined soybean oil. To our knowledge, this has never been reported before and represents an important accomplishment regarding the traceability of genetically modified organisms in refined oils."

According to Thomas Redick, detection and prevention of cross-pollination is possible through the suggestions offered by the Farm Service Agency (FSA) and Natural Resources Conservation Service (NRCS). Suggestions include educating farmers on the importance of coexistence, providing farmers with tools and incentives to promote coexistence, conduct research to understand and monitor gene flow, provide assurance of quality and diversity in crops, provide compensation for actual economic losses for farmers.

Controversies

The genetically modified foods controversy consists of a set of disputes over the use of food made from genetically modified crops. The disputes involve consumers, farmers, biotechnology companies, governmental regulators, non-governmental organizations, environmental and political activists and scientists. The major disagreements include whether GM foods can be safely consumed, harm the environment and/or are adequately tested and regulated. The objectivity of scientific research and publications has been challenged. Farming-related disputes include the use and impact of pesticides, seed production and use, side effects on non-GMO crops/farms, and potential control of the GM food supply by seed companies.

The conflicts have continued since GM foods were invented. They have occupied the media, the courts, local, regional and national governments and international organizations.

Genetically modified food controversies are disputes over the use of foods and other goods derived from genetically modified crops instead of conventional crops, and other uses of genetic engineering in food production. The dispute involves consumers, farmers, biotechnology companies, governmental regulators, non-governmental organizations, and scientists. The key areas of controversy related to genetically modified food (GM food or GMO food) are whether such food should be labeled, the role of government regulators, the objectivity of scientific research and publication, the effect of genetically modified crops on health and the environment, the effect on pesticide resistance, the impact of such crops for farmers, and the role of the crops in feeding the world population. In addition, products derived from GMO organisms play a role in the production of ethanol fuels and pharmaceuticals.

Specific concerns include mixing of genetically modified and non-genetically modified products in the food supply, effects of GMOs on the environment, the rigor of the regulatory process, and consolidation of control of the food supply in companies that make and sell GMOs. Advocacy groups such as the Center for Food Safety, Organic Consumers Association, Union of Concerned Scientists, and Greenpeace, say risks have not been adequately identified and managed, and they have questioned the objectivity of regulatory authorities.

The safety assessment of genetically engineered food products by regulatory bodies starts with an evaluation of whether or not the food is *substantially equivalent* to non-genetically engineered counterparts that are already deemed fit for human consumption. No reports of ill effects have been documented in the human population from genetically modified food. There is a scientific consensus that currently available food derived from GM crops poses no greater risk to human health than conventional food, but that each GM food needs to be tested on a case-by-case basis before introduction. Nonetheless, members of the public are much less likely than scientists to perceive GM foods as safe. The legal and regulatory status of GM foods varies by country, with some nations banning or restricting them, and others permitting them with widely differing degrees of regulation.

Public Perception

Consumer concerns about food quality first became prominent long before the advent of GM foods in the 1990s. Upton Sinclair's novel *The Jungle* led to the 1906 Pure Food and Drug Act, the first major US legislation on the subject. This began an enduring concern over the purity and later "naturalness" of food that evolved from a focus on sanitation to include added ingredients such as preservatives and flavors and sweeteners, residues such as pesticides, the rise of organic food as a category and finally to concerns over GM food. The public came to see the latter as "unnatural" which created a reverse halo effect.

Specific perceptions include genetic engineering as meddling with naturally evolved biological processes, scientific limitations on comprehending potential negative ramifications. An opposing perception is that genetic engineering is itself an evolution of traditional selective breeding.

Surveys indicate public concerns that eating genetically modified food is harmful, that biotechnology is risky, that more information is needed and that consumers need control over whether to take such risks. A diffuse sense that social and technological change is accelerating and that people cannot affect this change context becomes focused when such changes affect food. Leaders in driving public perception of the harms of such food in the media include Jeffrey M. Smith, Dr. Oz, Oprah, and Bill Maher; organizations include Organic Consumers Association, Greenpeace (especially with regard to Golden rice) and Union of Concerned Scientists.

Religious groups have raised concerns over whether genetically modified food will remain kosher or halal. In 2001 no such foods had been designated as unacceptable by Orthodox rabbis or Muslim leaders. However, some Jewish groups dispute this designation.

Environmental groups such as Friends of the Earth, include genetic engineering in general as an environmental and political concern. Other groups such as GMWatch and The Institute of Science in Society concentrate mostly or solely on opposing genetically modified crops.

Food writer Michael Pollan does not oppose eating genetically modified foods, but expressed concerns about biotechnology companies holding the intellectual property of the foods people depend on, and about the effects of the growing corporatization of large-scale agriculture. To address these problems, Pollan has brought up the idea of open sourcing GM foods. The idea has since been adopted to varying degrees by companies like Syngenta, and is being promoted by organizations such as the New America Foundation. Some organizations, like The BioBricks Foundation, have already worked out open-source licenses that could prove useful in this endeavour.

Reviews and Polls

A 2003 *EMBO Reports* article reported that the *Public Perceptions of Agricultural Biotechnologies in Europe* project (PABE) found the public neither accepting nor rejecting GMOs. Instead, PABE found that public had "key questions" about GMOs: "Why do we need GMOs? Who benefits from their use? Who decided that they should be developed and how? Why were we not better informed about their use in our food, before their arrival on the market? Why are we not given an effective choice about whether or not to buy these products? Have potential long-term and irreversible consequences been seriously evaluated, and by whom? Do regulatory authorities have sufficient powers to effectively regulate large companies? Who wishes to develop these products? Can controls imposed by regulatory authorities be applied effectively? Who will be accountable in cases of unforeseen harm?" PABE also found that the public's scientific knowledge does not control public opinion, since scientific facts do not answer these questions. PABE also found that the public does not demand "zero risk" in GM food discussions and is "perfectly aware that their lives are full of risks that need to be counterbalanced against each other and against the potential benefits. Rather than zero risk, what they demanded was a more realistic assessment of risks by regulatory authorities and GMO producers."

In 2006, the Pew Initiative on Food and Biotechnology made public a review of U.S. survey results from 2001-2006. The review showed that Americans' knowledge of GM foods and animals was low throughout the period. During this period protests against Calgene's Flavr Savr GM tomato mistakenly described it as containing fish genes, confusing it with DNA Plant Technology's fish tomato experimental transgenic organism, which was never commercialized.

A 2007 survey by the Food Standards Australia New Zealand found that in Australia, where labeling is mandatory, 27% of Australians checked product labels to see whether GM ingredients were present when initially purchasing a food item.

A 2009 review article of European consumer polls concluded that opposition to GMOs in Europe has been gradually decreasing, and that about 80% of respondents did not "actively avoid GM products when shopping". The 2010 "Eurobarometer" survey, which assesses public attitudes about biotech and the life sciences, found that cisgenics, GM crops made from plants that are crossable by conventional breeding, evokes a smaller reaction than transgenic methods, using genes from species that are taxonomically very different.

A 2010 Deloitte survey found that 34% of U.S. consumers were very or extremely concerned about GM food, a 3% reduction from 2008. The same survey found gender differences: 10% of men were extremely concerned, compared with 16% of women, and 16% of women were unconcerned, compared with 27% of men.

A 2013 poll by *The New York Times* showed that 93% of Americans wanted labeling of GM food.

The 2013 vote rejecting Washington State's GM food labeling I-522 referendum came shortly after the 2013 World Food Prize was awarded to employees of Monsanto and Syngenta. The award has drawn criticism from opponents of genetically modified crops.

With respect to the question of "Whether GMO foods were safe to eat," the gap between the opinion of the public and that of American Association for the Advancement of Science scientists is very wide with 88% of AAAS scientists saying yes in contrast to 37% of the general public.

Public Relations Campaigns and Protests

Anti-GMO and anti-Monsanto protests in Washington, D.C.

March Against Monsanto in Stockholm, Sweden, May 2013.

In May 2012, a group called "Take the Flour Back" led by Gerald Miles protested plans by a group from Rothamsted Experimental Station, based in Harpenden, Hertfordshire, England, to conduct an experimental trial wheat genetically modified to repel aphids. The researchers, led by John Pickett, wrote a letter to the group in early May 2012, asking them to call off their protest, aimed for 27 May 2012. Group member Lucy Harrap said that the group was concerned about spread of the crops into nature, and cited examples of outcomes in the United States and Canada. Rothamsted Research and Sense About Science ran question and answer sessions about such a potential.

The March Against Monsanto is an international grassroots movement and protest against Monsanto corporation, a producer of genetically modified organism (GMOs) and Roundup, a glyphosate-based herbicide. The movement was founded by Tami Canal in response to the failure of California Proposition 37, a ballot initiative which would have required labeling food products made from GMOs. Advocates support mandatory labeling laws for food made from GMOs.

The initial march took place on May 25, 2013. The number of protesters who took part is uncertain; figures of "hundreds of thousands" and the organizers' estimate of "two million" were variously cited. Events took place in between 330 and 436 cities around the world, mostly in the United States. Many protests occurred in Southern California, and some participants carried signs expressing support for mandatory labeling of GMOs that read "Label GMOs, It's Our Right to Know", and "Real Food 4 Real People". Canal said that the movement would continue its "anti-GMO cause" beyond the initial event. Further marches occurred in October 2013 and in May 2014 and 2015. The protests were reported by news outlets including ABC News, the Associated Press, *The Washington Post*, *The Los Angeles Times*, *USA Today*, and CNN (in the United States), and *Russia Today* and *The Guardian* (outside the United States).

Monsanto said that it respected people's rights to express their opinion on the topic, but maintained that its seeds improved agriculture by helping farmers produce more from their land while conserving resources, such as water and energy. The company reiterated that genetically modified foods were safe and improved crop yields. Similar sentiments were expressed by the Hawaii Crop Improvement Association, of which Monsanto is a member.

In July 2013, the agricultural biotechnology industry launched a GMO transparency initiative called GMO Answers to address consumers' questions about GM foods in the U.S. food supply. GMO Answers' resources included conventional and organic farmers, agribusiness experts, scientists, academics, medical doctors and nutritionists, and "company experts" from founding members of the Council for Biotechnology Information, which funds the initiative. Founding members include BASF, Bayer CropScience, Dow AgroSciences, DuPont, Monsanto Company and Syngenta.

In October 2013, a group called The European Network of Scientists for Social and Environmental Responsibility (ENSSER), posted a statement claiming that there is no scientific consensus on the safety of GMOs, which was signed by about 200 scientists in various fields in its first week. On January 25, 2015, their statement was formally published as a whitepaper by Environmental Sciences Europe.

Direct Action

Earth Liberation Front, Greenpeace and others have disrupted GMO research around the world. Within the UK and other European countries, as of 2014 80 crop trials by academic or governmental research institutes had been destroyed by protesters. In some cases, threats and violence against people or property were carried out. In 1999, activists burned the biotech lab of Michigan State University, destroying the results of years of work and property worth $400,000.

In 1987, the ice-minus strain of *P. syringae* became the first genetically modified organism (GMO) to be released into the environment when a strawberry field in California was sprayed with the bacteria. This was followed by the spraying of a crop of potato seedlings. The plants in both test fields were uprooted by activist groups, but were re-planted the next day.

In 2011, Greenpeace paid reparations when its members broke into the premises of an Australian scientific research organization, CSIRO, and destroyed a genetically modified wheat plot. The sentencing judge accused Greenpeace of cynically using junior members to avoid risking their own freedom. The offenders were given 9-month suspended sentences.

On August 8, 2013 protesters uprooted an experimental plot of golden rice in the Philippines. British author, journalist, and environmental activist Mark Lynas reported in *Slate* that the vandalism was carried out by a group led by the extreme-left KMP, to the dismay of other protesters. Golden rice is designed prevent vitamin A deficiency which, according to Helen Keller International, blinds or kills hundreds of thousands of children annually in developing countries.

Conspiracy Theories

There are various conspiracy theories related to the production and sale of genetically modified crops and genetically modified food that have been identified by some commentators such as Michael Shermer. Generally, these conspiracy theories posit that GMOs are being knowingly and maliciously introduced into the food supply either as a means to unduly enrich agribusinesses or as a means to poison or pacify the population.

A work seeking to explore risk perception over GMOs in Turkey identified a belief among the conservative political and religious figures who were opposed to GMOs that GMOs were "a conspiracy by Jewish Multinational Companies and Israel for world domination." Additionally, a Latvian study showed that a segment of the population believed that GMOs were part of a greater conspiracy theory to poison the population of the country.

Lawsuits

Foundation on Economic Trends v. Heckler

In 1983, environmental groups and protesters delayed the field tests of the genetically modified ice-minus strain of *P. syringae* with legal challenges.

Alliance for Bio-Integrity v. Shalala

In this case, the plaintiff argued both for mandatory labeling on the basis of consumer demand, and that GMO foods should undergo the same testing requirements as food additives because they are "materially changed" and have potentially unidentified health risks. The plaintiff also alleged that the FDA did not follow the Administrative Procedures Act in formulating and disseminating its policy on GMO's. The federal district court rejected all of those arguments and found that the FDA's determination that GMO's are Generally Recognized as Safe was neither arbitrary nor capricious. The court gave deference to the FDA's process on all issues, leaving future plaintiffs little legal recourse to challenge the FDA's policy on GMO's.

Diamond v. Chakrabarty

This case was on the question of whether GMOs can be patented.

Scientific Publishing

Scientific publishing on the safety and effects of GM foods is controversial.

Bt Maize

One of the first incidents occurred in 1999, when *Nature* published a paper on potential toxic effects of Bt maize on butterflies. The paper produced a public uproar and demonstrations, however by 2001 multiple follow-up studies had concluded that "the most common types of Bt maize pollen are not toxic to monarch larvae in concentrations the insects would encounter in the fields" and that they had "brought that particular question to a close."

Concerned scientists began to patrol the scientific literature and react strongly, both publicly and privately, to discredit conclusions they view as flawed in order to prevent unjustified public outcry and regulatory action. A 2013 *Scientific American* article noted that a "tiny minority" of biologists have published concerns about GM food, and said that scientists who support the use of GMOs in food production are often overly dismissive of them.

Restrictive End-User Agreements

Prior to 2010, scientists wishing to conduct research on commercial GM plants or seeds were unable to do so, because of restrictive end-user agreements. Cornell University's Elson Shields was the spokesperson for one group of scientists who opposed such restrictions. The group submitted a statement to the United States Environmental Protection Agency (EPA) in 2009 protesting that "as a result of restrictive access, no truly independent research can be legally conducted on many critical questions regarding the technology".

A 2009 *Scientific American* editorial quoted a scientist who said that several studies that were initially approved by seed companies were blocked from publication when they returned "unflattering" results. While favoring protection of intellectual property rights, the editors called for the restrictions to be lifted and for the EPA to require, as a condition of approval, that independent researchers have unfettered access to genetically modified products for research.

In December 2009, the American Seed Trade Association agreed to "allow public researchers greater freedom to study the effects of GM food crops." The companies signed blanket agreements permitting such research. This agreement left many scientists optimistic about the future, However, some scientists still express concern as to whether this agreement has the ability to "alter what has been a research environment rife with obstruction and suspicion." Monsanto previously had research agreements (i.e., Academic Research Licenses) with approximately 100 universities that allowed for university scientists to conduct research on their GM products with no oversight.

Reviews

A 2011 analysis by Diels *et al.* reviewed 94 peer-reviewed studies pertaining to GMO safety to assess whether conflicts of interest correlated with outcomes that cast GMOs in a favorable light. They found that financial conflict of interest was not associated with study outcome ($p = 0.631$) while author affiliation to industry (i.e., a professional conflict of interest) was strongly associated with study outcome ($p < 0.001$). Of the 94 studies that were analyzed, 52% did not declare funding. 10% of the studies were categorized as "undetermined" with regard to professional conflict of interest. Of the 43 studies with financial or professional conflicts of interest, 28 studies were compositional studies. According to Marc Brazeau, an association between professional conflict of interest and positive study outcomes can be skewed because companies typically contract with independent researchers to perform follow-up studies only after in-house research uncovers favorable results. In-house research that uncovers negative or unfavorable results for a novel GMO is generally not further pursued.

A 2013 review of 1,783 papers on genetically modified crops and food published between 2002 and 2012 found no plausible evidence of dangers from the use of then marketed GM crops. Biofortified, an independent nonprofit organization devoted to providing factual information and fostering discussion about agriculture, especially plant genetics and genetic engineering, planned to add the studies found by the Italian group to its database of studies about GM crops, GENERA.

In a 2014 review, Zdziarski *et al.* examined published studies of the histopathology of GI tracts of rats that were fed diets derived from GM crops, and identified some systemic flaws in this area of the scientific literature. Most studies were performed years after the approval of the crop for human consumption. Papers were often imprecise in their descriptions of the histological results and the selection of study endpoints, and lacked necessary details about methods and results. The authors called for the development of better study guidelines for determining the long-term safety of eating GM foods.

A 2016 study by the US National Academies of Sciences, Engineering, and Medicine concluded that GM foods are safe for human consumption and they could find no conclusive evidence that they harm the environment nor wildlife. They analysed over 1.000 studies over the previous 30 years that GM crops have been available, reviewed 700 written presentations submitted by interested bodies and heard 80 witnesses. They concluded that GM crops had given farmers economic advantages but found no evidence that GM crops had increased yields. They also noted that weed resistance to GM crops could cause major agricultural problems but this could be addressed by better farming procedures.

Alleged Data Manipulation

A University of Naples investigation suggested that images in eight papers on animals were intentionally altered and/or misused. The leader of the research group, Federico Infascelli, rejected the claim. The research concluded that mother goats fed GM soybean meal secreted fragments of the foreign gene in their milk. In December 2015 one of the papers was retracted for "self-plagiarism", although the journal noted that the results remained valid. A second paper was retracted in March 2016 after The University of Naples concluded that "multiple heterogeneities were likely attributable to digital manipulation, raising serious doubts on the reliability of the findings".

Health

There is a scientific consensus that currently available food derived from GM crops poses no greater risk to human health than conventional food, but that each GM food needs to be tested on a case-by-case basis before introduction. Nonetheless, members of the public are much less likely than scientists to perceive GM foods as safe. The legal and regulatory status of GM foods varies by country, with some nations banning or restricting them, and others permitting them with widely differing degrees of regulation.

The ENTRANSFOOD project was a European Commission-funded scientist group chartered to set a research program to address public concerns about the safety and value of agricultural biotechnology. It concluded that "the combination of existing test methods provides a sound test-regime to assess the safety of GM crops." In 2010, the European Commission Directorate-General for Research and Innovation reported that "The main conclusion to be drawn from the efforts of more than 130 research projects, covering a period of more than 25 years of involving more than 500 independent research groups, is that biotechnology, and in particular GMOs, are not per se more risky than e.g. conventional plant breeding technologies."

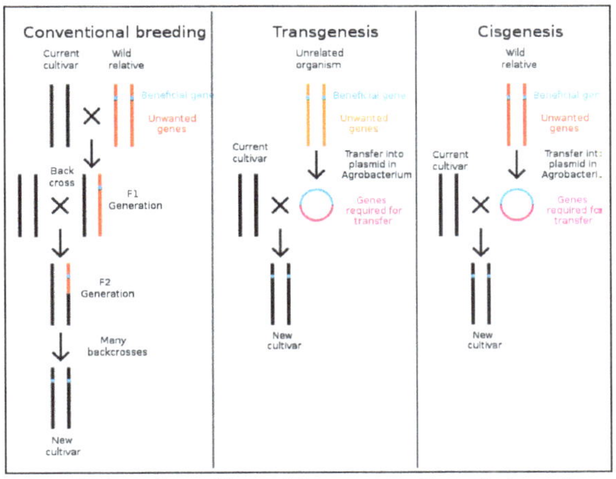

Comparison of conventional plant breeding with transgenic and cisgenic genetic modification.

Consensus among scientists and regulators pointed to the need for improved testing technologies and protocols. Transgenic and cisgenic organisms are treated similarly when assessed. However, in 2012 the European Food Safety Authority (EFSA) GMO Panel said that "novel hazards" could be associated with transgenic strains. In a 2016 review, Domingo concluded that studies in recent years had established that GM soybeans, rice, corn, and wheat do not differ from the corresponding conventional crops in terms of short-term human health effects, but recommended that further studies of long-term effects be conducted.

Substantial Equivalence

Most conventional agricultural products are the products of genetic manipulation via traditional cross-breeding and hybridization.

Governments manage the marketing and release of GM foods on a case-by-case basis. Countries differ in their risk assessments and regulations. Marked differences distinguish the US from Europe. Crops not intended as foods are generally not reviewed for food safety. GM foods are not tested in humans before marketing because they are not a single chemical, nor are they intended to be ingested using specific doses and intervals, which complicate clinical study design. Regulators examine the genetic modification, related protein products and any changes that those proteins make to the food.

Regulators check that GM foods are "substantially equivalent" to their conventional counterparts, to detect any negative unintended consequences. New protein(s) that differ from conventional food proteins or anomalies that arise in the substantial equivalence comparison require further toxicological analysis.

"The World Health Organization, the American Medical Association, the U.S. National Academy of Sciences, the British Royal Society, and every other respected organization that has examined the evidence has come to the same conclusion: consuming foods containing ingredients derived from GM crops is no riskier than consuming the same foods containing ingredients from crop plants modified by conventional plant improvement techniques."

–American Association for the Advancement of Science

In 1999, Andrew Chesson of the Rowett Research Institute warned that substantial equivalence testing "could be flawed in some cases" and that current safety tests could allow harmful substances to enter the human food supply. The same year Millstone, Brunner and Mayer argued that the standard was a pseudo-scientific product of politics and lobbying that was created to reassure consumers and aid biotechnology companies to reduce the time and cost of safety testing. They suggested that GM foods have extensive biological, toxicological and immunological tests and that substantial equivalence should be abandoned. This commentary was criticized for misrepresenting history, for distorting existing data and poor logic. Kuiper claimed that it oversimplified safety assessments and that equivalence testing involves more than chemical tests, possibly including toxicity testing. Keler and Lappe supported Congressional legislation to replace the substantial equivalence standard with safety studies. In a 2016 review, Domingo criticized the use of the "substantial equivalence" concept as a measure of the safety of GM crops.

Kuiper examined this process further in 2002, finding that substantial equivalence does not measure absolute risks, but instead identifies differences between new and existing products. He claimed that characterizing differences is properly a starting point for a safety assessment and "the concept of substantial equivalence is an adequate tool in order to identify safety issues related to genetically modified products that have a traditional counterpart". Kuiper noted practical difficulties in applying this standard, including the fact that traditional foods contain many toxic or carcinogenic chemicals and that existing diets were never proven to be safe. This lack of knowledge re conventional food means that modified foods may differ in anti-nutrients and natural toxins that have never been identified in the original plant, possibly allowing harmful changes to be missed. In turn, positive modifications may also be missed. For example, corn damaged by insects often contains high levels of fumonisins, carcinogenic toxins made by fungi that travel on insects' backs and that grow in the wounds of damaged corn. Studies show that most Bt corn has lower levels of fumonisins than conventional insect-damaged corn. Workshops and consultations organized by the OECD, WHO, and FAO have worked to acquire data and develop better understanding of conventional foods, for use in assessing GM foods.

A survey of publications comparing the intrinsic qualities of modified and conventional crop lines (examining genomes, proteomes and metabolomes) concluded that GM crops had less impact on gene expression or on protein and metabolite levels than the variability generated by conventional breeding.

In a 2013 review Herman (Dow AgroSciences) and Price (FDA, retired) argued that transgenesis is less disruptive than traditional breeding techniques because the latter routinely involve more changes (mutations, deletions, insertions and rearrangements) than the relatively limited changes (often single gene) in genetic engineering. The FDA found that all of the 148 transgenic events that they evaluated to be substantially equivalent to their conventional counterparts, as have Japanese regulators for 189 submissions including combined-trait products. This equivalence was confirmed by more than 80 peer-reviewed publications. Hence, the authors argue, compositional equivalence studies uniquely required for GM food crops may no longer be justified on the basis of scientific uncertainty.

Allergenicity

A well-known risk of genetic modification is the introduction of an allergen. Allergen testing is routine for products intended for food, and passing those tests is part of the regulatory requirements. Organizations such as the European Green Party and Greenpeace emphasize this risk. A 2005 review of the results from allergen testing stated that "no biotech proteins in foods have been documented to cause allergic reactions". Regulatory authorities require that new modified foods be tested for allergenicity before they are marketed.

GMO proponents note that because of the safety testing requirements, the risk of introducing a plant variety with a new allergen or toxin is much smaller than from traditional breeding processes, which do not require such tests. Genetic engineering can have less impact on the expression of genomes or on protein and metabolite levels than conventional breeding or (non-directed) plant mutagenesis. Toxicologists note that "conventional food is not risk-free; allergies occur with many known and even new conventional foods. For example, the kiwi fruit was introduced into the U.S. and the European markets in the 1960s with no known human allergies; however, today there are people allergic to this fruit."

Genetic modification can also be used to remove allergens from foods, potentially reducing the risk of food allergies. A hypo-allergenic strain of soybean was tested in 2003 and shown to lack the major allergen that is found in the beans. A similar approach has been tried in ryegrass, which produces pollen that is a major cause of hay fever: here a fertile GM grass was produced that lacked the main pollen allergen, demonstrating that hypoallergenic grass is also possible.

The development of genetically modified products found to cause allergic reactions has been halted by the companies developing them before they were brought to market. In the early 1990s, Pioneer Hi-Bred attempted to improve the nutrition content of soybeans intended for animal feed by adding a gene from the Brazil nut. Because they knew that people have allergies to nuts, Pioneer ran in vitro and skin prick allergy tests. The tests showed that the transgenic soy was allergenic. Pioneer Hi-Bred therefore discontinued further development. In 2005, a pest-resistant field pea developed by the Australian Commonwealth Scientific and Industrial Research Organisation for use as a pasture crop was shown to cause an allergic reaction in mice. Work on this variety was immediately halted. These cases have been used as evidence that genetic modification can produce unexpected and dangerous changes in foods, and as evidence that safety tests effectively protect the food supply.

During the Starlink corn recalls in 2000, a variety of GM maize containing the *Bacillus thuringiensis*

(Bt) protein Cry9C, was found contaminating corn products in U.S. supermarkets and restaurants. It was also found in Japan and South Korea. Starlink corn had only been approved for animal feed as the Cry9C protein lasts longer in the digestive system than other Bt proteins raising concerns about its potential allergenicity. In 2000, Taco Bell-branded taco shells sold in supermarkets were found to contain Starlink, resulting in a recall of those products, and eventually led to the recall of over 300 products. Sales of StarLink seed were discontinued and the registration for the Starlink varieties was voluntarily withdrawn by Aventis in October 2000. Aid sent by the United Nations and the United States to Central African nations was also found to be contaminated with StarLink corn and the aid was rejected. The U.S. corn supply has been monitored for Starlink Bt proteins since 2001 and no positive samples have been found since 2004. In response, GeneWatch UK and Greenpeace set up the GM Contamination Register in 2005. During the recall, the United States Centers for Disease Control evaluated reports of allergic reactions to StarLink corn, and determined that no allergic reactions to the corn had occurred.

Horizontal Gene Transfer

Horizontal gene transfer is the movement of genes from one organism to another in a manner other than reproduction.

The risk of horizontal gene transfer between GMO plants and animals is very low and in most cases is expected to be lower than background rates. Two studies on the possible effects of feeding animals with genetically modified food found no residues of recombinant DNA or novel proteins in any organ or tissue samples. Studies found DNA from the M13 virus, Green fluorescent protein and Rubisco genes in the blood and tissue of animals, and in 2012, a paper suggested that a specific microRNA from rice could be found at very low quantities in human and animal serum. Other studies however, found no or negligible transfer of plant microRNAs into the blood of humans or any of three model organisms.

Another concern is that the antibiotic resistance gene commonly used as a genetic marker in transgenic crops could be transferred to harmful bacteria, creating resistant superbugs. A 2004 study involving human volunteers examined whether the transgene from modified soy would transfer to bacteria that live in the human gut. As of 2012 it was the only human feeding study to have been conducted with GM food. The transgene was detected in three volunteers from a group of seven who had previously had their large intestines removed for medical reasons. As this gene transfer did not increase after the consumption of the modified soy, the researchers concluded that gene transfer did not occur. In volunteers with intact digestive tracts, the transgene did not survive. The antibiotic resistance genes used in genetic engineering are naturally found in many pathogens and antibiotics these genes confer resistance to are not widely prescribed.

Animal Feeding Studies

Reviews of animal feeding studies mostly found no effects. A 2014 review found that the performance of animals fed GM feed was similar to that of animals fed "isogenic non-GE crop lines". A 2012 review of 12 long-term studies and 12 multigenerational studies conducted by public research laboratories concluded that none had discovered any safety problems linked to consumption of GM food. A 2009 review by Magaña-Gómez found that although most studies concluded that modified foods do not differ in nutrition or cause toxic effects in animals, some did report adverse changes at a cellular level caused by specific modified foods. The review concluded that

"More scientific effort and investigation is needed to ensure that consumption of GM foods is not likely to provoke any form of health problem". Dona and Arvanitoyannis' 2009 review concluded that "results of most studies with GM foods indicate that they may cause some common toxic effects such as hepatic, pancreatic, renal, or reproductive effects and may alter the hematological, biochemical, and immunologic parameters". Reactions to this review in 2009 and 2010 noted that Dona and Arvanitoyannis had concentrated on articles with an anti-modification bias that were refuted in peer-reviewed articles elsewhere. Flachowsky concluded in a 2005 review that food with a one-gene modification were similar in nutrition and safety to non-modified foods, but he noted that food with multiple gene modifications would be more difficult to test and would require further animal studies. A 2004 review of animal feeding trials by Aumaitre and others found no differences among animals eating genetically modified plants.

In 2007, Domingo's search of the Pubmed database using 12 search terms indicated that the "number of references" on the safety of GM or transgenic crops was "surprisingly limited," and he questioned whether the safety of GM food had been demonstrated. The review also stated that its conclusions were in agreement with three earlier reviews. However, Vain found 692 research studies in 2007 that focused on GM crop and food safety and found increasing publication rates of such articles in recent years. Vain commented that the multidisciplinarian nature of GM research complicated the retrieval of studies based on it and required many search terms (he used more than 300) and multiple databases. Domingo and Bordonaba reviewed the literature again in 2011 and said that, although there had been a substantial increase in the number of studies since 2006, most were conducted by biotechnology companies "responsible of commercializing these GM plants."

Human Studies

While some groups and individuals have called for more human testing of GM food, multiple obstacles complicate such studies. The General Accounting Office (in a review of FDA procedures requested by Congress) and a working group of the Food and Agricultural and World Health organizations both said that long-term human studies of the effect of GM food are not feasible. The reasons included lack of a plausible hypothesis to test, lack of knowledge about the potential long-term effects of conventional foods, variability in the ways humans react to foods and that epidemiological studies were unlikely to differentiate modified from conventional foods, which come with their own suite of unhealthy characteristics.

Additionally, ethical concerns guide human subject research. These mandate that each tested intervention must have a potential benefit for the human subjects, such as treatment for a disease or nutritional benefit (ruling out, e.g., human toxicity testing). Kimber claimed that the "ethical and technical constraints of conducting human trials, and the necessity of doing so, is a subject that requires considerable attention." Food with nutritional benefits may escape this objection. E.g., GM rice has been tested for nutritional benefits, namely, increased levels of Vitamin A.

Controversial Studies

Pusztai Affair

Arpad Pusztai published the first peer-reviewed paper to find negative effects from GM food consumption in 1999. Pusztai fed rats potatoes transformed with the *Galanthus nivalis* agglutinin

(GNA) gene from the *Galanthus* (snowdrop) plant, allowing the tuber to synthesise the GNA lectin protein. While some companies were considering growing GM crops expressing lectin, GNA was an unlikely candidate. Lectin is toxic, especially to gut epithelia. Pusztai reported significant differences in the thickness of the gut epithelium, but no differences in growth or immune system function.

On June 22, 1998, an interview on Granada Television's current affairs programme World in Action, Pusztai said that rats fed on the potatoes had stunted growth and a repressed immune system. A media frenzy resulted. Pusztai was suspended from the Rowett Institute. Misconduct procedures were used to seize his data and ban him from speaking publicly. The Rowett Institute and the Royal Society reviewed his work and concluded that the data did not support his conclusions. The work was criticized on the grounds that the unmodified potatoes were not a fair control diet and that any rat fed only potatoes would suffer from protein deficiency. Pusztai responded by stating that all diets had the same protein and energy content and that the food intake of all rats was the same.

Bt Corn

A 2011 study was the first to evaluate the correlation between maternal and fetal exposure to Bt toxin produced in GM maize and to determine exposure levels of the pesticides and their metabolites. It reported the presence of pesticides associated with the modified foods in women and in pregnant women's fetuses. The paper and related media reports were criticized for overstating the results. Food Standards Australia New Zealand (FSANZ) posted a direct response, saying that the suitability of the ELISA method for detecting the Cry1Ab protein was not validated and that no evidence showed that GM food was the protein's source. The organization also suggested that even had the protein been detected its source was more likely conventional or organic food.

Séralini Affair

In 2007, 2009, and 2011, Gilles-Éric Séralini published re-analysis studies that used data from Monsanto rat-feeding experiments for three modified maize varieties (insect-resistant MON 863 and MON 810 and glyphosate-resistant NK603). He concluded that the data showed liver, kidney and heart damage. The European Food Safety Authority (EFSA) then concluded that the differences were all within the normal range. EFSA also stated that Séralini's statistics were faulty. EFSA's conclusions were supported by FSANZ, a panel of expert toxicologists, and the French High Council of Biotechnologies Scientific Committee (HCB).

In 2012, Séralini's lab published a paper that considered the long-term effects of feeding rats various levels of GM glyphosate-resistant maize, conventional glyphosate-treated maize, and a mixture of the two strains. The paper concluded that rats fed the modified maize had severe health problems, including liver and kidney damage and large tumors. The study provoked widespread criticism. Séralini held a press conference just before the paper was released in which he announced the release of a book and a movie. He allowed reporters to have access to the paper before his press conference only if they signed a confidentiality agreement under which they could not report other scientists' responses to the paper. The press conference resulted in media coverage emphasizing a connection between GMOs, glyphosate, and cancer. Séralini›s publicity stunt yielded criticism from other scientists for prohibiting critical commentary. Criticisms included insufficient statistical power and that Séralini's Sprague-Dawley rats were inappropriate for a lifetime study (as opposed to a shorter toxicity study) because of their tendency to develop cancer (one study found that more than 80% normally got cancer). The Organisation for Economic Co-operation and Development guidelines recommended using 65 rats per experiment instead of the 10 in Séralini's. Other criticisms included the lack of data regarding food amounts and specimen growth rates, the lack of a dose–response relationship (females fed three times the standard dose showed a decreased number of tumours) and no identified mechanism for the tumour increases. Six French national academies of science issued an unprecedented joint statement condemning the study and the journal that published it. *Food and Chemical Toxicology* published many critical letters, with only a few expressing support. National food safety and regulatory agencies also reviewed the paper and dismissed it. In March 2013, Séralini responded to these criticisms in the same journal that originally published his study, and a few scientists supported his work. In November 2013, the editors of *Food and Chemical Toxicology* retracted the paper. The retraction was met with protests from Séralini and his supporters. In 2014, the study was republished by a different journal, *Environmental Sciences Europe*, in an expanded form, including the raw data that Séralini had originally refused to reveal.

Environment

Genetically modified crops are planted in fields much like regular crops. There they interact directly with organisms that feed on the crops and indirectly with other organisms in the food chain. The pollen from the plants is distributed in the environment like that of any other crop. This distribution has led to concerns over the effects of GM crops on the environment. Potential effects include gene flow, pesticide resistance and greenhouse gas emissions.

Non-Target Organisms

A major use of GM crops is in insect control through the expression of the *cry* (crystal delta-endotoxins) and *Vip* (vegetative insecticidal proteins) genes from *Bacillus thuringiensis* (Bt). Such toxins could affect other insects in addition to targeted pests such as the European corn borer. Bt proteins have been used as organic sprays for insect control in France since 1938 and the US since 1958, with no reported ill effects. *Cry* proteins selectively target Lepidopterans (moths and butterflies). As a toxic mechanism, *cry* proteins bind to specific receptors on the membranes of mid-gut (epithelial) cells, resulting in their rupture. Any organism that lacks the appropriate receptors in its gut is unaffected by the *cry* protein, and therefore is not affected by Bt. Regulatory agencies assess the potential for transgenic plants to affect non-target organisms before approving their commercial release.

In 1999 a paper stated that, in a laboratory environment, pollen from Bt maize dusted onto milkweed could harm the monarch butterfly. A collaborative research exercise over the following two years by several groups of scientists in the US and Canada studied the effects of Bt pollen in both the field and the laboratory. The study resulted in a risk assessment concluding that any risk posed to butterfly populations was negligible. A 2002 review of the scientific literature concluded that "the commercial large-scale cultivation of current Bt–maize hybrids did not pose a significant risk to the monarch population" and noted that despite large-scale planting of genetically modified crops, the butterfly's population was increasing. However, the herbicide glyphosate used to grow

GMOs kills milkweed, the only food source of Monarch butterflies, and by 2015 about 90% of the U.S. population has declined.

Lövei et al. analyzed laboratory settings and found that Bt toxins could affect non-target organisms, generally closely related to the intended targets. Typically, exposure occurs through the consumption of plant parts, such as pollen or plant debris, or through Bt ingestion by predators. A group of academic scientists criticized the analysis, writing: “We are deeply concerned about the inappropriate methods used in their paper, the lack of ecological context, and the authors’ advocacy of how laboratory studies on non-target arthropods should be conducted and interpreted”.

Biodiversity

Crop genetic diversity might decrease due to the development of superior GM strains that crowd others out of the market. Indirect effects might affect other organisms. To the extent that agrochemicals impact biodiversity, modifications that increase their use, either because successful strains require them or because the accompanying development of resistance will require increased amounts of chemicals to offset increased resistance in target organisms.

Studies comparing the genetic diversity of cotton found that in the US diversity has either increased or stayed the same, while in India it has declined. This difference was attributed to the larger number of modified varieties in the US compared to India. A review of the effects of Bt crops on soil ecosystems found that in general they “appear to have no consistent, significant, and long-term effects on the microbiota and their activities in soil”.

The diversity and number of weed populations has been shown to decrease in farm-scale trials in the United Kingdom and in Denmark when comparing herbicide-resistant crops to their conventional counterparts. The UK trial suggested that the diversity of birds could be adversely affected by the decrease in weed seeds available for foraging. Published farm data involved in the trials showed that seed-eating birds were more abundant on conventional maize after the application of the herbicide, but that there were no significant differences in any other crop or prior to herbicide treatment. A 2012 study found a correlation between the reduction of milkweed in farms that grew glyphosate-resistant crops and the decline in adult monarch butterfly populations in Mexico. *The New York Times* reported that the study “raises the somewhat radical notion that perhaps weeds on farms should be protected.

A 2005 study designed to “simulate the impact of a direct overspray on a wetland” with four different agrochemicals (carbaryl (Sevin), malathion, 2,4-dichlorophenoxyacetic acid, and glyphosate in a Roundup formulation) by creating artificial ecosystems in tanks and then applying “each chemical at the manufacturer’s maximum recommended application rates” found that “species richness was reduced by 15% with Sevin, 30% with malathion, and 22% with Roundup, whereas 2,4-D had no effect”. The study has been used by environmental groups to argue that use of agrochemicals causes unintended harm to the environment and to biodiversity.

Secondary Pests

Several studies documented surges in secondary pests within a few years of adoption of Bt cotton. In China, the main problem has been with mirids, which have in some cases “completely eroded all benefits from Bt cotton cultivation". A 2009 study in China concluded that the increase in secondary pests depended on local temperature and rainfall conditions and occurred in half the villages studied. The increase in insecticide use for the control of these secondary insects was far smaller than the reduction in total insecticide use due to Bt cotton adoption. A 2011 study based on a survey of 1,000 randomly selected farm households in five provinces in China found that the reduction in pesticide use in Bt cotton cultivars was significantly lower than that reported in research elsewhere: The finding was consistent with a hypothesis that more pesticide sprayings are needed over time to control emerging secondary pests, such as aphids, spider mites, and lygus bugs. Similar problems have been reported in India, with mealy bugs and aphids.

Gene Flow

Genes from a GMO may pass to another organism just like an endogenous gene. The process is known as outcrossing and can occur in any new open-pollinated crop variety. Introduced traits potentially can cross into neighboring plants of the same or closely related species through three different types of gene flow: crop-to-crop, crop-to-weedy, and crop-to-wild. In crop-to-crop, genetic information from a genetically modified crop is transferred to a non-genetically modified crop. Crop-to-weedy transfer refers to the transfer of genetically modified material to a weed, and crop-to-wild indicates transfer from a genetically modified crop to a wild, undomesticated plant and/or crop. There are concerns that the spread of genes from modified organisms to unmodified relatives could produce species of weeds resistant to herbicides that could contaminate nearby non-genetically modified crops, or could disrupt the ecosystem, This is primarily a concern if the transgenic organism has a significant survival capacity and can increase in frequency and persist in natural populations. This process, whereby genes are transferred from GMOs to wild relatives, is different from the development of so-called "superweeds" or "superbugs" that develop resistance to pesticides under natural selection.

In most countries environmental studies are required before approval of a GMO for commercial purposes, and a monitoring plan must be presented to identify unanticipated gene flow effects.

In 2004, Chilcutt and Tabashnik found Bt protein in kernels of in a refuge (a conventional crop planted to harbor pests that might otherwise become resistant a pesticide associated with the GMO) implying that gene flow had occurred.

In 2005, scientists at the UK Centre for Ecology and Hydrology reported the first evidence of horizontal gene transfer of pesticide resistance to weeds, in a few plants from a single season; they found no evidence that any of the hybrids had survived in subsequent seasons.

In 2007, the U.S. Department of Agriculture fined Scotts Miracle-Gro $500,000 when modified DNA from GM creeping bentgrass, was found within relatives of the same genus (*Agrostis*) as well as in native grasses up to 21 km (13 mi) from the test sites, released when freshly cut, wind-blown grass.

In 2009 Mexico created a regulatory pathway for GM maize, but because Mexico is maize's center of diversity, concerns were raised about GM maize's effects on local strains. A 2001 report found

Bt maize cross-breeding with conventional maize in Mexico. The data in this paper was later described as originating from an artifact and the publishing journal *Nature* stated that "the evidence available is not sufficient to justify the publication of the original paper", although it did not retract the paper. A subsequent large-scale study, in 2005, found no evidence of gene flow in Oaxaca. However, other authors claimed to have found evidence of such gene flow.

A 2010 study showed that about 83 percent of wild or weedy canola tested contained genetically modified herbicide resistance genes. According to the researchers, the lack of reports in the United States suggested that oversight and monitoring were inadequate. A 2010 report stated that the advent of glyphosate-resistant weeds could cause GM crops to lose their effectiveness unless farmers combined glyphosate with other weed-management strategies.

One way to avoid environmental contamination is genetic use restriction technology (GURT), also called "Terminator". This uncommercialized technology would allow the production of crops with sterile seeds, which would prevent the escape of GM traits. Groups concerned about food supplies had expressed concern that the technology would be used to limit access to fertile seeds. Another hypothetical technology known as "Traitor" or "T-GURT", would not render seeds sterile, but instead would require application of a chemical to GM crops to activate engineered traits. Groups such as Rural Advancement Foundation International raised concerns that further food safety and environmental testing needed to be done before T-GURT would be commercialized.

Escape of Modified Crops

The escape of genetically modified seed into neighboring fields, and the mixing of harvested products, is of concern to farmers who sell to countries that do not allow GMO imports.

In 1999 scientists in Thailand claimed they had discovered unapproved glyphosate-resistant GM wheat in a grain shipment, even though it was only grown in test plots. No mechanism for the escape was identified.

In 2000, Aventis StarLink GM corn was found in US markets and restaurants. It became the subject of a recall that started when Taco Bell-branded taco shells sold in supermarkets were found to contain it. StarLink was then discontinued. Registration for Starlink varieties was voluntarily withdrawn by Aventis in October 2000.

American rice exports to Europe were interrupted in 2006 when the LibertyLink modification was found in commercial rice crops, although it had not been approved for release. An investigation by the USDA's Animal and Plant Health Inspection Service (APHIS) failed to determine the cause of the contamination.

In May 2013, unapproved glyphosate-resistant GM wheat (but that had been approved for human consumption) was discovered in a farm in Oregon in a field that had been planted with winter wheat. The strain was developed by Monsanto, and had been field-tested from 1998 to 2005. The discovery threatened US wheat exports which totaled $8.1 billion in 2012. Japan, South Korea and Taiwan temporarily suspended winter wheat purchases as a result of the discovery. As of August 30, 2013, while the source of the modified wheat remained unknown, Japan, South Korea and Taiwan had resumed placing orders.

Coexistence with Conventional Crops

The US has no legislation governing the relationship among mixtures of farms that grow organic, conventional and GM crops. The country relies on a "complex but relaxed" combination of three federal agencies (FDA, EPA, and USDA/APHIS) and states' common law tort systems to manage coexistence. The Secretary of Agriculture convened an Advisory Committee on Biotechnology and 21st Century Agriculture (AC21) to study coexistence and make recommendations about the issue. The members of AC21 included representatives of the biotechnology industry, the organic food industry, farming communities, the seed industry, food manufacturers, State governments, consumer and community development groups, the medical profession, and academic researchers. AC21 recommended that a study assess the potential for economic losses to US organic farmers; that any serious losses lead to a crop insurance program, an education program to ensure that organic farmers put appropriate contracts in place and that neighboring GMO farmers take appropriate containment measures. Overall the report supported a diverse agriculture system supporting diverse farming systems.

The EU implemented regulations specifically governing co-existence and traceability. Traceability has become commonplace in the food and feed supply chains of most countries, but GMO traceability is more challenging given strict legal thresholds for unwanted mixing. Since 2001, conventional and organic food and feedstuffs can contain up to 0.9% of authorised modified material without carrying a GMO label. (any trace of non-authorised modification is cause for a shipment to be rejected). Authorities require the ability to trace, detect and identify GMOs, and the several countries and interested parties created a non-governmental organization, Co-Extra, to develop such methods.

Chemical Use

Birth defects in Hawaii caused a controversy, and public outcry of people concerned over heavy pesticide usage. Hawaii uses ten times the amount of pesticides compared to the rest of the U.S.

Pesticides

Pesticides destroy, repel or mitigate pests (an organism that attacks or competes with a crop). A 2014 meta-analysis covering 147 original studies of farm surveys and field trials, and 15 studies from the researchers conducting the study, concluded that adoption of GM technology had reduced chemical pesticide use by 37%, with the effect larger for insect-tolerant crops than herbicide-tolerant crops.

Herbicides

The development of glyphosate-tolerant (Roundup Ready) plants changed the herbicide use profile away from more persistent, higher toxicity herbicides, such as atrazine, metribuzin and alachlor, and reduced the volume and harm of herbicide runoff. A study by Chuck Benbrook concluded that the spread of glyphosate-resistant weeds had increased US herbicide use. That study cited a 23% increase (.3 kilograms/hectare) for soybeans from 1996-2006, a 43% (.9 kg/ha) increase for cotton from 1996-2010 and a 16% (.5 kg/ha) decrease for corn from 1996-2010. However, this study came under scrutiny because Benbrook did not consider the fact that glyphosate is less toxic than other herbicides, thus net toxicity may decrease even as use increases. Graham Brookes accused Benbrook of subjective herbicide estimates because his data, provided by the National Agricultural Statistics Service, does not distinguish between genetically modified and non-genetically modified crops. Brookes had earlier published a study that found that the use of biotech crops had reduced the volume and environmental impact of herbicide and other pesticides, which contradicted Benbrook. Brookes stated that Benbrook had made "biased and inaccurate" assumptions.

Insecticides

A claimed environmental benefit of Bt-cotton and maize is reduced insecticide use. A PG Economics study concluded that global pesticide use was reduced by 286,000 tons in 2006, decreasing pesticidal environmental impact by 15%. A survey of small Indian farms between 2002 and 2008 concluded that Bt cotton adoption had led to higher yields and lower pesticide use. Another study concluded that insecticide use on cotton and corn during the years 1996 to 2005 fell by 35,600,000 kilograms (78,500,000 lb) of active ingredient, roughly equal to the annual amount applied in the European Union. A Bt cotton study in six northern Chinese provinces from 1990 to 2010 concluded that it halved the use of pesticides and doubled the level of ladybirds, lacewings and spiders and extended environmental benefits to neighbouring crops of maize, peanuts and soybeans.

Resistant Insect Pests

Resistance evolves naturally after a population has been subjected to selection pressure via repeated use of a single pesticide. In November 2009, Monsanto scientists found that the pink bollworm had become resistant to first generation Bt cotton in parts of Gujarat, India—that generation expresses one Bt gene, *Cry1Ac*. This was the first instance of Bt resistance confirmed by Monsanto. Similar resistance was later identified in Australia, China, Spain and the US.

One strategy to delay Bt-resistance is to plant pest refuges using conventional crops, thereby diluting any resistant genes. Another is to develop crops with multiple Bt genes that target different receptors within the insect. In 2012, a Florida field trial demonstrated that army worms were resistant to Dupont-Dow's GM corn. This resistance was discovered in Puerto Rico in 2006, prompting Dow and DuPont to stop selling the product there. The European corn borer, one of Bt's primary targets, is also capable of developing resistance.

Economy

GM food's economic value to farmers is one of its major benefits, including in developing nations. A 2010 study found that Bt corn provided economic benefits of $6.9 billion over the previous 14 years in five Midwestern states. The majority ($4.3 billion) accrued to farmers producing non-Bt corn. This was attributed to European corn borer populations reduced by exposure to Bt corn, leaving fewer to attack conventional corn nearby. Agriculture economists calculated that "world surplus [increased by] $240.3 million for 1996. Of this total, the largest share (59%) went to U.S. farmers. Seed company Monsanto received the next largest share (21%), followed by US consumers (9%), the rest of the world (6%), and the germplasm supplier, Delta and Pine Land Company (5%)." PG Economics comprehensive 2012 study concluded that GM crops increased farm incomes worldwide by $14 billion in 2010, with over half this total going to farmers in developing countries.

The main Bt crop grown by small farmers in developing countries is cotton. A 2006 review of Bt cotton findings by agricultural economists concluded, “the overall balance sheet, though promising, is mixed. Economic returns are highly variable over years, farm type, and geographical location”. However, environmental activist Mark Lynas said that complete rejection of genetic engineering is “illogical and potentially harmful to the interests of poorer peoples and the environment”.

In 2013 the European Academies Science Advisory Council (EASAC) asked the EU to allow the development of agricultural GM technologies to enable more sustainable agriculture, by employing fewer land, water and nutrient resources. EASAC also criticizes the EU’s “timeconsuming and expensive regulatory framework” and said that the EU had fallen behind in the adoption of GM technologies.

Industrial Agriculture

GM crops play a key role in intensive crop farming, which involves monoculture, use of herbicides and pesticides, use of equipment requiring large amounts of fuel and irrigation. Opponents such as Jonathan Latham of the *Bioscience Resource Project* and Vandana Shiva treat industrial agriculture and modified crops as closely related topics, and call for agriculture to adopt practices drastically reducing disruptions to the environment.

Proponents of conventional agriculture point to its high yields, low prices, and wider choices and claim that technology is necessary to feed a growing world population.

Developing Nations

Disagreements about developing nations include the claimed need for increased food supplies, and how to achieve such an increase. Some scientists suggest that a second Green Revolution including use of modified crops is needed to provide sufficient food. The potential for genetically modified food to help developing nations was recognised by the International Assessment of Agricultural Science and Technology for Development, but as of 2008 they had found no conclusive evidence of a solution.

Skeptics such as John Avise claim that apparent shortages are caused by problems in food distribution and politics, rather than production. Other critics say that the world has so many people because the second green revolution adopted unsustainable agricultural practices that left the world with more mouths to feed than the planet can sustain. Pfeiffer claimed that even if technological farming could feed the current population, its dependence on fossil fuels, which in 2006 he incorrectly predicted would reach peak output in 2010, would lead to a catastrophic rise in energy and food prices.

Claimed deployment constraints to developing nations include the lack of easy access, equipment costs and intellectual property rights that hurt developing countries. The Consultative Group on International Agricultural Research (CGIAR), an aid and research organization, was praised by the World Bank for its efforts, but the bank recommended that they shift to genetics research and productivity enhancement. Obstacles include access to patents, commercial licenses and the difficulty that developing countries have in accessing genetic resources and other intellectual property. The International Treaty on Plant Genetic Resources for Food and Agriculture attempted to remedy this problem, but results have been inconsistent. As a result, "orphan crops", such as teff, millets, cowpeas and indigenous plants, which are important in these countries receive little investment.

Writing about Norman Borlaug's 2000 publication *Ending world hunger: the promise of biotechnology and the threat of antiscience zealotry*, the authors argued that Borlaug's warnings were still true in 2010:

GM crops are as natural and safe as today's bread wheat, opined Dr. Borlaug, who also reminded agricultural scientists of their moral obligation to stand up to the antiscience crowd and warn policy makers that global food insecurity will not disappear without this new technology and ignoring this reality would make future solutions all the more difficult to achieve.

Yield

US maize yields were flat until the 1930s, when the adoption of conventional hybrid seeds caused them to increase by ~.8 bushels/acre (1937-1955). Thereafter a combination of improved genetics, fertilizer and pesticide availability and mechanization raised the rate of increase to 1.9 bushels per acre per year. In the years since the advent of GM maize, the rate increased slightly to 2.0. Average US maize yields were 174.2 bushels per acre in 2014.

Commercial GM crops have traits that reduce yield loss from insect pressure or weed interference.

2014 Review

A 2014 review concluded that GM crops' effects on farming were positive. According to *The Economist*, the meta-analysis considered all published English-language examinations of the agronomic and economic impacts between 1995 and March 2014. The study found that herbicide-tolerant crops have lower production costs, while for insect-resistant crops the reduced pesticide use was offset by higher seed prices, leaving overall production costs about the same.

Yields increased 9% for herbicide tolerance and 25% for insect resistance. Farmers who adopted GM crops made 69% higher profits than those who did not. The review found that GM crops help farmers in developing countries, increasing yields by 14 percentage points.

The researchers considered some studies that were not peer-reviewed, and a few that did not report sample sizes. They attempted to correct for publication bias, by considering sources beyond academic journals. The large data set allowed the study to control for potentially confounding variables such as fertiliser use. Separately, they concluded that the funding source did not influence study results.

2010 Review

A 2010 article supported by CropLife International summarised the results of 49 peer reviewed studies. On average, farmers in developed countries increased yields by 6% and 29% in developing countries.

Tillage decreased by 25–58% on herbicide-resistant soybeans. Glyphosate-resistant crops allowed farmers to plant rows closer together as they did not have to control post-emergent weeds with mechanical tillage. Insecticide applications on Bt crops were reduced by 14–76%. 72% of farmers worldwide experienced positive economic results.

2009 Review

In 2009 the Union of Concerned Scientists, a group opposed to genetic engineering and cloning of food animals, summarized peer-reviewed studies on the yield contribution of GM soybeans and maize in the US. The report concluded that other agricultural methods had made a greater contribution to national crop yield increases in recent years than genetic engineering.

Wisconsin Study

A study unusually published as correspondence rather than as an article examined maize modified to express four traits (resistance to European corn borer, resistance to corn root worm, glyphosate tolerance and glyfosinate tolerance) singly and in combination in Wisconsin fields from 1990-2010. The variance in yield from year to year was reduced, equivalent to a yield increase of 0.8-4.2 bushels per acre. Bushel per acre yield changes were +6.4 for European corn borer resistance, +5.76 for glufosinate tolerance, -5.98 for glyphosate tolerance and -12.22 for corn rootworm resistance. The study found interactions among the genes in multi-trait hybrid strains, such that the net effect varied from the sum of the individual effects. For example, the combination of European corn borer resistance and glufosinate tolerance increased yields by 3.13, smaller than either of the individual traits.

Market Dynamics

The seed industry is dominated by a small number of vertically integrated firms. In 2011, 73% of the global market was controlled by 10 companies.

In 2001, the USDA reported that industry consolidation led to economies of scale, but noted that the move by some companies to divest their seed operations questioned the long-term viability of these conglomerates. Two economists have said that the seed companies' market power could raise welfare despite their pricing strategies, because "even though price discrimination is often considered to be an unwanted market distortion, it may increase total welfare by increasing total output and by making goods available to markets where they would not appear otherwise."

Market share gives firms the ability to set or influence price, dictate terms, and act as a barrier to entry. It also gives firms bargaining power over governments in policy making. In March 2010, the US Department of Justice and the US Department of Agriculture held a meeting in Ankeny, Iowa, to look at the competitive dynamics in the seed industry. Christine Varney, who heads the antitrust division in the Justice Department, said that her team was investigating whether biotech-seed patents were being abused. A key issue was how Monsanto licenses its patented glyphosate-tolerance trait that was in 93 percent of US soybeans grown in 2009. About 250 family farmers, consumers and other critics of corporate agriculture held a town meeting prior to the government meeting to protest Monsanto's purchase of independent seed companies, patenting seeds and then raising seed prices.

Intellectual Property

Traditionally, farmers in all nations saved their own seed from year to year. However, since the early 1900s hybrid crops have been widely used in the developed world and seeds to grow these crops are purchased each year from seed producers. The offspring of the hybrid corn, while still viable, lose hybrid vigor (the beneficial traits of the parents). This benefit of first-generation hybrid seeds is the primary reason for not planting second-generation seed. However, for non-hybrid GM crops, such as GM soybeans, seed companies use intellectual property law and tangible property common law, each expressed in contracts, to prevent farmers from planting saved seed. For example, Monsanto's typical bailment license (covering transfer of the seeds themselves) forbids saving seeds, and also requires purchasers to sign a separate patent license agreement.

Corporations say that they need to prevent seed piracy, to fulfill financial obligations to shareholders, and to finance further development. DuPont spent approximately half its $2 billion research and development (R&D) budget on agriculture in 2011 while Monsanto spends 9-10% of sales on R&D.

Detractors such as Greenpeace say that patent rights give corporations excessive control over agriculture. The Center for Ecoliteracy claimed that "patenting seeds gives companies excessive power over something that is vital for everyone." A 2000 report stated, "If the rights to these tools are strongly and universally enforced - and not extensively licensed or provided pro bono in the developing world - then the potential applications of GM technologies described previously are unlikely to benefit the less developed nations of the world for a long time" (i.e. until after the restrictions expire).

Monsanto has patented its seed and it obligates farmers who choose to buy its seeds to sign a license agreement, obligating them store or sell, but not plant, all the crops that they grow.

Lawsuits Filed Against Farmers for Patent Infringement

Monsanto has filed patent infringement suits against 145 farmers, but proceeded to trial with only 11. In some of the latter, the defendants claimed unintentional contamination by gene flow, but Monsanto won every case. Monsanto Canada's Director of Public Affairs stated, "It is not, nor has it ever been Monsanto Canada's policy to enforce its patent on Roundup Ready crops when they are present on a farmer's field by accident...Only when there has been a knowing and deliberate violation of its patent rights will Monsanto act." In 2009 Monsanto announced that after its soybean patent expires in 2014, it will no longer prohibit farmers from planting soybean seeds that they grow.

One example of such litigation is the Monsanto v. Schmeiser case. This case is widely misunderstood. In 1997, Percy Schmeiser, a canola breeder and grower in Bruno, Saskatchewan, discovered that one of his fields had canola that was resistant to Roundup. He had not purchased this seed, which had blown onto his land from neighboring fields. He later harvested the area and saved the crop in the back of a pickup truck.Before the 1998 planting, Monsanto representatives informed Schmeiser that using this crop for seed would infringe the patent, and offered him a license, which Schmeiser refused.According to the Canadian Supreme Court, after this conversation "Schmeiser nevertheless took the harvest he had saved in the pick-up truck to a seed treatment plant and had it treated for use as seed. Once treated, it could be put to no other use. Mr. Schmeiser planted the treated seed in nine fields, covering approximately 1,000 acres in all....A series of independent tests by different experts confirmed that the canola Mr. Schmeiser planted and grew in 1998 was

95 to 98 percent Roundup resistant."After further negotiations between Schmeiser and Monsanto broke down, Monsanto sued Schmeiser for patent infringement and prevailed in the initial case. Schmeiser appealed and lost, and appealed again to the Canadian Supreme Court, which in 2004 ruled 5 to 4 in Monsanto's favor, stating that "it is clear on the findings of the trial judge that the appellants saved, planted, harvested and sold the crop from plants containing the gene and plant cell patented by Monsanto."

International Trade

GM crops have been the source of international trade disputes and tensions within food-exporting nations over whether introduction of genetically modified crops would endanger exports to other countries.

In Canada in 2010, flax exports to Europe were rejected when traces of an experimental GM flax were found in shipments. This led a member of Parliament to propose Private Member's Bill C-474, which would have required that "an analysis of potential harm to export markets be conducted before the sale of any new genetically engineered seed is permitted." Opponents claimed that "incorporating stringent socio-economic standards into the science-based regulatory system could spell the end of private research funding; because if private biotechnology companies can't see the possibility of a return on their investment, they'll invest their research budget elsewhere." The bill was defeated in 2011, in a "vote of 176 to 97."

Regulation

Labeling

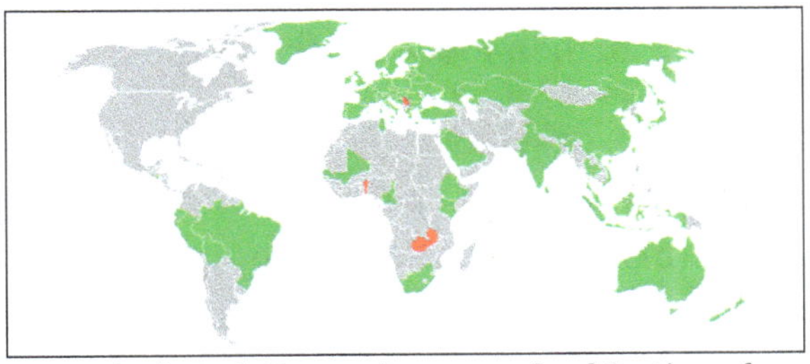

Green: Mandatory labeling required; Red: Ban on import and cultivation of genetically engineered food. Regulations are pending in Israel and the United States.

In 2014, 64 countries required labeling of all GM foods. These include the European Union, Japan, Australia, New Zealand, Russia, China and India. As of March 2015, Israel was in the process of issuing regulations for labeling of food with ingredients from GMOs.

Alaska required labeling of GMO fish and shellfish in 2005, even though no GM fish had been approved by the FDA at the time. A 2014 Vermont law went into effect on July 1, 2016, and some food manufacturers (including General Mills, Mars, Kellogg's, the Campbell Soup Company, PepsiCo, ConAgra, Frito-Lay, and Bimbo Bakeries USA) began distributing products either locally or nation-wide with labels such as "Partially produced with Genetic Engineering". Other manufacturers removed about 3,000 non-compliant products from sale in Vermont. The federal government of the United States passed a law at the end of that month pre-empting all state laws, including Vermont's. The law requires labeling regulations to be issued by July 2018, and allows indirect disclosure such as with a phone number, bar code, or web site. It is unclear whether the rules will require labeling of oils and sugars from GM crops, where the final product does not contain any "genetic material" as mentioned in the law.

Prior to the new federal rules taking effect, while it does require pre-market approval, the U.S. Food and Drug Administration has not required GMO labeling as long as there are no differences in health, environmental safety, and consumer expectations based on the packaging. The federal rules come after GMO labeling was debated in many state legislatures and defeated in popular referendums in Oregon (2002 and 2014), Colorado (2014), California Proposition 37 (2012), and Washington Initiative 522 (2012). Connecticut and Maine had passed laws in 2013 and 2014 respectively, which would have required GMO food labels if Northeast states with a population of at least 20 million had passed similar laws (and for Connecticut, representing at least four states).

Other jurisdictions make such labeling voluntary or have had plans to require labeling. Major GM food crop exporters like the United States (until 2018), Argentina, and Canada have adopted voluntary labeling approaches; China and Brazil have major GM (largely non-food) crops and have adopted mandatory labelling.

Arguments

The American Public Health Association, the British Medical Association and the Public Health Association of Australia support mandatory labeling. The European Commission argued that mandatory labeling and traceability are needed to allow for informed choice, avoid potential misleading of consumers and facilitate the withdrawal of products if adverse effects on health or the environment are discovered. A 2007 study on the effect of labeling laws found that once labeling went into effect, few products continued to contain GM ingredients. The study also found that costs were higher in food-exporting than in food-importing countries.

The American Medical Association (AMA) and the American Association for the Advancement of Science have opposed mandatory labeling absent scientific evidence of harm. The AMA said that even voluntary labeling is misleading unless accompanied by focused consumer education. The AAAS stated that mandatory labeling "can only serve to mislead and falsely alarm consumers".

[Labeling] efforts are not driven by evidence that GM foods are actually dangerous. Indeed, the science is quite clear: crop improvement by the modern molecular techniques of biotechnology is safe. Rather, these initiatives are driven by a variety of factors, ranging from the persistent perception that such foods are somehow "unnatural" and potentially dangerous to the desire to gain competitive advantage by legislating attachment of a label meant to alarm. Another misconception used as a rationale for labeling is that GM crops are untested.

Objectivity of Regulatory Bodies

Groups such as the Union of Concerned Scientists and Center for Food Safety that have expressed concerns about the FDA's lack of a requirement for additional testing for GMO's, lack of required labeling and the presumption that GMO's are "Generally Recognized as Safe" (GRAS), have questioned whether the FDA is too close to companies that seek approval for their products.

Critics in the U.S. protested the appointment of lobbyists to senior positions in the Food and Drug Administration. Michael R. Taylor, a former Monsanto lobbyist, was appointed as a senior adviser to the FDA on food safety in 1991. After leaving the FDA, Taylor became a vice-president of Monsanto. On 7 July 2009, Taylor returned to government as a senior adviser to the FDA Commissioner.

In 2001, when the Starlink corn recall became public, the U.S. Environmental Protection Agency was criticized for being slow to react by Joseph Mendelson III of the Center for Food Safety. He also criticized the EPA and Aventis CropScience for statements at the time of the recall, that indicated they did not anticipate that such a thing would happen.

The Canadian Biotechnology Advisory Committee that reviewed Canada's regulations in 2003 was accused by environmental and citizen groups of not representing the full spectrum of public interests and for being too closely aligned to industry groups.

Most of the Chinese National Biosafety Committee are involved in biotechnology, a situation that led to criticisms that they do not represent a wide enough range of public concerns.

Litigation and Regulation Disputes

United States

Four federal district court suits have been brought against Animal and Plant Health Inspection Service (APHIS), the agency within USDA that regulates genetically modified plants. Two involved field trials (herbicide-tolerant turfgrass in Oregon; pharmaceutical-producing corn and sugar in Hawaii) and two the deregulation of GM alfalfa. and GM sugar beet. APHIS lost all four cases at trial, with the judges ruling they failed to diligently follow the guidelines set out in the National Environmental Policy Act. However, the Supreme Court overturned the nationwide ban on GM alfalfa and an appeal court allowed the partial deregulation of GM sugar beets. After APHIS prepared Environmental Impact Statements for both alfalfa and sugar beets they were approved.

In 2014, Maui County, Hawaii approved an initiative calling for a moratorium on GMO production and research. The initiative specified penalties including fines and jail for knowing violations and did not limit its scope to commercial agriculture. The initiative passed by about 50.2 to 47.9 percent.

On December 15, 2015, the New York Times ran an op-ed titled "Are You Eating Frankenfish?", saying that the United States congress will debate whether genetically engineered salmon should be labeled.

European Union

Until the 1990s, Europe's regulation was less strict than in the U.S. In 1998, the use of MON810, a Bt expressing maize conferring resistance to the European corn borer, was approved for commercial cultivation in Europe. However, in the 1990s a series of unrelated food crises created consumer apprehension about food safety in general and eroded public trust in government oversight. A bovine spongiform encephalopathy outbreak was the most publicized. In 1998, a *de facto* moratorium led to the suspension of approvals of new GMOs in the EU pending the adoption of revised rules.

In the mid-1990s, government approval of some GMO crops in the United States precipitated public concern in Europe and led to a dramatic decrease in American exports to Europe. "Prior to 1997, corn exports to Europe represented about 4% of total US corn exports, generating about $300 million in sales....For example, before 1997, the U.S. sold about 1.75 million tons of corn annually to Spain and Portugal....But in the 1998–99 crop year, Spain bought less than a tenth of the previous year's amount and Portugal bought none at all."

In May 2003, the US and twelve other countries filed a formal complaint with the World Trade Organization that the EU was violating international trade agreements, by blocking imports of US farm products through its ban on GM food. The countries argued that the EU's regulatory process was far too slow and its standards were unreasonable given the scientific evidence showing that the crops were safe. The case was lobbied by Monsanto and France's Aventis, as well as by US agricultural groups such as the National Corn Growers Association. In response, in June 2003, the European Parliament ratified a U.N. biosafety protocol regulating international trade in GM food, and in July agreed to new regulations requiring labeling and traceability, as well as an opt-out provision for individual countries. The approval of new GMOs resumed in May 2004. While GMOs have been approved since then, approvals remain controversial and various countries have utilized opt-out provisions. In 2006, the World Trade Organization ruled that the pre-2004 restrictions had been violations, although the ruling had little immediate effect since the moratorium had already been lifted.

In late 2007, the US ambassador to France recommended "moving to retaliation" to cause "some pain" against France and the European Union in an attempt to fight the French ban and changes in European policy toward genetically modified crops, according to a US government diplomatic cable obtained by WikiLeaks.

20 out of 28 European Countries (including Switzerland) said No to GMOs until October 2015.

These are the countries which have/implement a ban on GMO:

- Austria
- Belgium for the Wallonia region
- Britain for Scotland, Wales and Northern Ireland
- Bulgaria
- Croatia
- Cyprus
- Denmark
- France
- Germany (except for research)
- Greece
- Hungary

- Italy
- Latvia
- Lithuania
- Luxembourg
- Malta
- Netherlands
- Philippines
- Poland
- Slovenia
- Switzerland

Australia

In May 2014, the Supreme Court of the Australian state of Western Australia dismissed "Marsh v. Baxter". The plaintiff was Steve Marsh, an organic farmer, and the defendant was Michael Baxter, his lifelong neighbor, who grew GM canola. In late 2010 Marsh found seeds from Baxter's crop in his fields. Later, Marsh found escaped GM canola growing amidst his crop. Marsh reported the seed and plants to his local organic certification board, and lost the organic certification of some 70 per cent of his 478 hectare farm. Marsh sued on the grounds that Baxter used a method of harvesting his crop that was substandard and negligent, and on the basis that his land had been widely contaminated. In its summary judgment, the court found that approximately 245 cut canola plants were blown by the wind into Marsh's property, Eagle's Rest. However, Baxter's method (swathing) was "orthodox and well accepted harvest methodology". "In 2011, eight GM canola plants were found to have grown up as self-sown volunteer plants on Eagle Rest," which "were identified and pulled out", and "no more volunteer RR canola plants grew on Eagle Rest in subsequent years." The summary judgment stated that the loss of organic certification "was occasioned by the erroneous application of governing NASAA Standards applicable to NASAA organic operators as regards GMOs (genetically modified organisms) at the time." and that "[t]he absence of a reliable underlying evidentiary platform to support a perpetual injunction against swathing was a significant deficiency".

On June 18, 2014 Marsh announced that he had filed an appeal. One ground was the costs of $803,989 awarded against him. The appeal hearing commenced on 23 March 2015 and was adjourned on 25 March "to deal with an order to ascertain whether Mr Baxter's defence has been financially supported by GM-seed supplier Monsanto and/or the Pastoralists and Graziers Association (PGA)".

Philippines

A petition filed May 17, 2013 by environmental group Greenpeace Southeast Asia and farmer-scientist coalition Masipag (Magsasaka at Siyentipiko sa Pagpapaunlad ng Agrikultura) asked the appellate court to stop the planting of Bt eggplant in test fields, saying the impacts of such an undertaking to the environment, native crops and human health are still unknown. The Court of Appeals granted the petition, citing the precautionary principle stating "when human activities may lead to threats of serious and irreversible damage to the environment that is scientifically plausible but uncertain, actions shall be taken to avoid or diminish the threat." Respondents filed a motion for reconsideration in June 2013 and on September 20, 2013 the Court of Appeals chose to uphold their May decision saying the bt talong field trials violate the people's constitutional right to a "balanced and healthful ecology." The Supreme Court on Tuesday, December 8, 2015 permanently stopped the field testing for Bt (*Bacillus thuringiensis*) talong (eggplant), upholding the decision of the Court of Appeals which stopped the field trials for the genetically modified eggplant.

Innovation in Technology and Regulatory law

The first genetically modified crops were made with transgenic approaches, introducing foreign genes and sometimes using bacteria to transfer the genes. In the US, these foreign genetic elements placed the resulting plant under the jurisdiction of the USDA under the Plant Protection Act. However, as of 2010, newer genetic engineering technologies like genome editing have allowed scientists to modify plant genomes without adding foreign genes, thus escaping USDA regulation. Critics have called for regulation to be changed to keep up with changing technology.

Legislation

African Controversies

In 2002, in the midst of a famine, Zambia refused emergency food aid that contained food from genetically modified crops, based on the precautionary principle.

During a conference in the Ethiopian capital of Addis Ababa, Kingsley Amoako, Executive Secretary of the United Nations Economic Commission for Africa (UNECA), encouraged African nations to accept GM food and expressed dissatisfaction in the public's negative opinion of biotechnology.

Studies for Uganda showed that transgenic bananas had a high potential to reduce rural poverty but that urban consumers with a relatively higher income might reject them.

Critics claimed that shipment of US food to southern Africa was more about promoting the adoption of biotech crops in the region than about hunger. The US was supplying Africa with meals and support during a food crisis they were facing in the early 2000s. However, once some of the African countries realized that these shipments contained GM maize, they rejected the shipments and stopped releasing the food that had been sent to them. Critics accused the US of "exploiting the Southern African famine as a public relations tool". The U.S. countered these comments by saying that European nations were letting millions of Africans suffer from hunger and starvation because of "irrational fears over hypothetical and unproven risks". The US had a pre-GMO policy of shipping US crops as food aid, rather than buying crops in/near the countries that needed aid. The US policy was claimed to be more costly than Europe's.

Genetically modified food controversies in Ghana have been widespread since 2013.

Indian Controversies

India is an agrarian country with around 60% of its people depending directly or indirectly upon agriculture. From 1995 to 2013, a total of 296,438 farmers have killed themselves in India, or an average of 16,469 suicides per year. During the same period, about 9.5 million people died per year in India from other causes including malnutrition, diseases and suicides that were non-farming related, or about 171 million deaths from 1995 to 2013. Activists and scholars have offered a number of conflicting reasons for farmer suicides, such as monsoon failure, high debt burdens, genetically modified crops, government policies, public mental health, personal issues and family problems. There are also accusations of states reporting inaccurate data on farmer suicides.

In India, GM cotton yields in Maharashtra, Karnataka, and Tamil Nadu resulted in an average 42% increase in yield in 2002, the first year of commercial planting. A severe drought in Andhra Pradesh that year prevented any increase in yield, because the GM strain was not drought tolerant. Drought-tolerant variants were later developed. Driven by substantially reduced losses to insect predation, by 2011 88% of Indian cotton was modified. Though disputed, the economic and environmental benefits of GM cotton to farmers have been documented. A study from 2002 through 2008 on the economic impacts of Bt cotton in India, showed that Bt cotton increased yields, profits and living standards of smallholder farmers. However, recently cotton bollworm has been developing resistance to Bt cotton. Consequently, in 2012 Maharashtra banned Bt cotton and ordered an independent socioeconomic study of its use. Indian regulators cleared the Bt brinjal, a genetically modified eggplant, for commercialisation in October 2009. After opposition by some scientists, farmers and environmental groups, a moratorium was imposed on its release in February 2010 "for as long as it is needed to establish public trust and confidence".

As of 1 January 2013, all foods containing GMOs must be labelled. The Legal Metrology (Packaged Commodities) Rules, 2011 states that "every package containing the genetically modified food shall bear at the top of its principal display panel the letters 'GM.'" The rules apply to 19 products including biscuits, breads, cereals and pulses, and a few others. The law faced criticism from consumer rights activists as well as from the packaged-food industry; both sides had major concerns that no logistical framework or regulations had been established to guide the law's implementation and enforcement. On March 21, 2014, the Indian government revalidated 10 GM-based food crops and allowed field trials of GM food crops, including wheat, rice and maize.

References

- Final Report of the PABE research project (December 2001). "Public Perceptions of Agricultural Biotechnologies in Europe". Commission of European Communities. Retrieved 12, February 2020
- Bashshur, Ramona (February 2013). "FDA and Regulation of GMOs". American Bar Association. Retrieved 03, April 2020
- Séralini G (2012). Tous Cobayes !: OGM, pesticides et produits chimiques [All Guinea Pigs: GMOs, pesticides and chemicals] (in French). Editions Flammarion. ISBN 9782081262362
- Pollack M, Shaffer G (2009). When Cooperation Fails: the international law and politics of genetically modified foods. Oxford University Press. ISBN 978-0-19-956705-8
- Webster, TM; Grey, TL (2015). "Glyphosate-Resistant Palmer Amaranth (Amaranthus palmeri) Morphology, Growth, and Seed Production in Georgia.". Weed Science. 63 (1): 264–272. doi:10.1614/ws-d-14-00051.1
- "Questions & Answers on Food from Genetically Engineered Plants". US Food and Drug Administration. 22 Jun 2015. Retrieved 18, July 2020
- Cowan, Tadlock (18 Jun 2011). "Agricultural Biotechnology: Background and Recent Issues" (PDF). Congressional Research Service (Library of Congress). pp. 33–38. Retrieved 22, August 2020

- Chokshi, Niraj (9 May 2014). "Vermont just passed the nation's first GMO food labeling law. Now it prepares to get sued". The Washington Post. Retrieved 17, March 2020
- Commissioner, Office of the. "Press Announcements - FDA takes several actions involving genetically engineered plants and animals for food". www.fda.gov. Retrieved 18, January 2020

Permissions

All chapters in this book are published with permission under the Creative Commons Attribution Share Alike License or equivalent. Every chapter published in this book has been scrutinized by our experts. Their significance has been extensively debated. The topics covered herein carry significant information for a comprehensive understanding. They may even be implemented as practical applications or may be referred to as a beginning point for further studies.

We would like to thank the editorial team for lending their expertise to make the book truly unique. They have played a crucial role in the development of this book. Without their invaluable contributions this book wouldn't have been possible. They have made vital efforts to compile up to date information on the varied aspects of this subject to make this book a valuable addition to the collection of many professionals and students.

This book was conceptualized with the vision of imparting up-to-date and integrated information in this field. To ensure the same, a matchless editorial board was set up. Every individual on the board went through rigorous rounds of assessment to prove their worth. After which they invested a large part of their time researching and compiling the most relevant data for our readers.

The editorial board has been involved in producing this book since its inception. They have spent rigorous hours researching and exploring the diverse topics which have resulted in the successful publishing of this book. They have passed on their knowledge of decades through this book. To expedite this challenging task, the publisher supported the team at every step. A small team of assistant editors was also appointed to further simplify the editing procedure and attain best results for the readers.

Apart from the editorial board, the designing team has also invested a significant amount of their time in understanding the subject and creating the most relevant covers. They scrutinized every image to scout for the most suitable representation of the subject and create an appropriate cover for the book.

The publishing team has been an ardent support to the editorial, designing and production team. Their endless efforts to recruit the best for this project, has resulted in the accomplishment of this book. They are a veteran in the field of academics and their pool of knowledge is as vast as their experience in printing. Their expertise and guidance has proved useful at every step. Their uncompromising quality standards have made this book an exceptional effort. Their encouragement from time to time has been an inspiration for everyone.

The publisher and the editorial board hope that this book will prove to be a valuable piece of knowledge for students, practitioners and scholars across the globe.

Index

www.ingramcontent.com/pod-product-compliance
Lightning Source LLC
LaVergne TN
LVHW080917210926
848485LV00004B/667
9781639872817